AF368921

# Evaluation of Boulder Deposits Linked to Late Neogene Hurricane Events

# Evaluation of Boulder Deposits Linked to Late Neogene Hurricane Events

Editors

**Markes E. Johnson**
**Jorge Ledesma-Vázquez**

MDPI • Basel • Beijing • Wuhan • Barcelona • Belgrade • Manchester • Tokyo • Cluj • Tianjin

*Editors*
Markes E. Johnson
Williams College
USA

Jorge Ledesma-Vázquez
Universidad Autónoma de Baja
California
Mexico

*Editorial Office*
MDPI
St. Alban-Anlage 66
4052 Basel, Switzerland

This is a reprint of articles from the Special Issue published online in the open access journal *Journal of Marine Science and Engineering* (ISSN 2077-1312) (available at: https://www.mdpi.com/journal/jmse/special_issues/boulder_deposits).

For citation purposes, cite each article independently as indicated on the article page online and as indicated below:

LastName, A.A.; LastName, B.B.; LastName, C.C. Article Title. *Journal Name* **Year**, *Volume Number*, Page Range.

**ISBN 978-3-0365-2578-5 (Hbk)**
**ISBN 978-3-0365-2579-2 (PDF)**

Cover image courtesy of Markes E. Johnson

# Contents

# About the Editors

**Markes E. Johnson** (Charles L. MacMillan Professor of Natural Science, Emeritus)

Professor Johnson taught courses in historical geology, paleontology and stratigraphy in the Geosciences Department at Williams College over a 35-year career. He grew up in the Midwest, where dolomite bluffs along the Upper Mississippi River drew his attention as a hobbyist collecting marine fossils that had lived in a vast continental sea. His undergraduate education in geology concluded with a BA degree (1971) from the University of Iowa. His advanced training in paleoecology through the Department of Geophysical Sciences at the University of Chicago culminated in a PhD degree (1977). Since 1990, Prof. Johnson has made one or two annual trips to the Baja California peninsula and Mexico's Gulf of California to study coastal deposits related to the Pliocene Warm Period and later Pleistocene epochs when sea level and global temperatures were higher than today. Since 2009, he has remained active with studies regarding the Miocene to Pleistocene history of many North Atlantic islands, including those of the Cape Verde, Canary, Madeira, and Azores archipelagos. Prof. Johnson was the recipient of the 2011 Nelson Bushnell Prize for excellence in scholarship and teaching at Williams College. He is a frequent guest lecturer accompanying excursions sponsored by the Williams College Society of Alumni to Scandinavia, the Iberian Peninsula, the Galapagos Islands, and Mexico's Gulf of California.

**Jorge Ledesma-Vázquez** (Professor Emeritus, Área de Geologia, Facultad de Ciencias Marinas)

Professor Ledesma-Vázquez taught courses in oceanography and coastal sedimentology in the Facultad de Ciencias Marinas for a 37-year career. He holds a degree in engineering geology from the Instituto Politénco Nacional (IPN) in Mexico City (1977), a MA degree from San Diego State University (1991) and a PhD from the Universidad Autónoma de Baja California (2000). His earliest visit to Baja California occurred in 1975 as part of a field trip with classmates from IPN, for which he obtained funding through a personal appeal to the president of Mexico. He has co-led many research expeditions and field courses involving students from UABC, Williams College, and other institutions. Since 2010, he has participated in field studies regarding island groups in the Cape Verde, Canary, and Madeira archipelagos that share similarities with islands in the Gulf of California. Prof. Ledesma-Vázquez received a Distinguished Alumnus Award from San Diego State University in 2019.

Journal of
*Marine Science and Engineering*

*Editorial*

# Evaluation of Boulder Deposits Linked to Late Neogene Hurricane Events

Markes E. Johnson [1,*] and Jorge Ledesma-Vázquez [2]

[1]  Department of Geosciences, Williams College, Williamstown, MA 01267, USA
[2]  Facultad de Ciencias Marinas, Universidad Autónoma de Baja California,
     Ensenada 22800, Baja California, Mexico; ledesma@uabc.edu.mx
*   Correspondence: mjohnson@williams.edu; Tel.: +1-413-2329

**Citation:** Johnson, M.E.;
Ledesma-Vázquez, J. Evaluation of
Boulder Deposits Linked to Late
Neogene Hurricane Events. *J. Mar. Sci.
Eng.* **2021**, *9*, 1278. https://doi.org/
10.3390/jmse9111278

Received: 11 November 2021
Accepted: 15 November 2021
Published: 17 November 2021

**Publisher's Note:** MDPI stays neutral
with regard to jurisdictional claims in
published maps and institutional affiliations.

The Neogene is a globally recognized interval of geologic time that lasted from 23 until 1.8 million years ago, also divided into two geological periods: the Miocene and the Pliocene [1]. Eight contributions to this Special Issue are available for study and comparison, which span not only the Neogene but the last 1.8 million years of Earth history including the Pleistocene glacial and inter-glacial epochs followed by the Holocene beginning only 10,000 years ago. Like today, past hurricane events were responsible for the erosion of rocky shores due to the impact of storm waves and for the development of flood deposits due to heavy rainfall after big storms made landfall. The former had the potential to result in coastal boulder beds (CBDs) eroded from sea cliffs and the latter in coastal outwash deposits (CODs) derived from upland sources. Two key events are widely recorded during the Neogene, known as the Middle Miocene climatic optimum (about 15 million years ago) and subsequent Pliocene climatic optimum (between ~4.5 and 3.0 million years ago) [2]. During those peak times, the global average temperature and average sea level were higher compared to today. Major sea storms originating as subtropical depressions may have been more common on the high seas and more effective in making landfall over islands and continental shores. The survey conducted by Ruban (2019) [3] was the first contribution to this Special Issue. It reviews 21 studies almost evenly divided between the Miocene and Pliocene. All feature a mixture of boulder-size clasts exceeding 25.6 cm in diameter, although the original author's intent was not always explicit with the identification of a CBD related to storm activity. Indeed, some examples are attributed to a delta setting more consistent with a COD and others are interpreted as tsunami deposits. The overall relevance of the Special Issue is that the Neogene acts as a casement for a bridge head that connects the past to the present time with increased global warming inducive to expanded hurricane activity [2].

The co-editors of this Special Issue (SI) have a particular interest in Mexico's Baja California peninsula and adjacent Gulf of California (also known as the Sea of Cortez). Three contributions to the SI detail CBDs along the eastern seaboard of the peninsula within the Gulf of California [4–6]. Studies conducted on Isla San Luis Gonzaga in the upper or northern part of the Gulf [4], as well as at Puerto Escondido on the peninsula's central coast [5], relate to the erosion of andesite sea cliffs during the Holocene time. The study on Isla San Diego in the lower or southern part of the Gulf [6] deals with Pleistocene storm events that occurred approximately 125,000 years ago. A unique feature of these studies is the application of a triangular plotting system that takes into account the shape of eroded shore bounders, based on measurements in three dimensions. More perfectly spherical boulders plot near the apex of the triangle, and boulders that are more elongated in shape plot towards the lower, right corner of the triangle. Relatively few boulders derived from andesite rocky shores [4,5] plot as spherical in shape, whereas the trend toward elongated boulders is notably prominent. The same is generally true of boulders derived from granodiorite rocky shores [6], which formed from the slow cooling of magma deeper within the earth's crust when compared to surface flows associated with the regionally more common andesite. However, the sheeted exfoliation of granodiorite typical of its natural weathering appears to have resulted in similarly shaped boulders favoring an elongated form. Another aspect of Isla San Diego includes a tail of eroded boulders observed as extending underwater from the southern end of the island for a distance of nearly 1.5 km [6], believed to be formed under the influence of recent storm activity.

Three contributions to the SI [7–9] relate to island deposits in the North Atlantic Ocean. From the Azorean island of Santa Maria [7], comparisons are made among paired CBDs of Pliocene and Pleistocene origin exposed along the southern coast of that island. A Pleistocene CBD [8] is described from Gran Canaria in the Canary Islands. A Holocene CBD is analyzed from Leka Island at a high latitude off the coast of Norway. These Atlantic-island studies conform to the same procedures entailed in the analysis of boulder shapes as shown for the CBDs from Mexico's Gulf of California, and each pays attention to the key factor of rock density. Generally smaller in size, the cobbles and boulders at Støypet on Leka Island were eroded from low-grade chromite ore with a density of 3.32 g/cm$^3$. This reflects the highest-density rock type yet studied for its hydrological properties in a coastal setting. The basalt from the Canary and Azores registers between 2.8 and 3.0 g/cm$^3$. The density of granodiorite from Isla San Diego in the Gulf of California was found to be 2.5 g/cm$^3$ [6], whereas andesite samples from Puerto Escondido and Isla San Luis Gonzaga were found to register a density of 2.3 g/cm$^3$ or higher [4,5]. While it is routine to consider the rock density when studying the hydrology of boulders under wave attack from individual study localities, the data from multiple localities summarized in one place makes for interesting comparisons. The same amount of wave energy necessary to budge the smallest boulder at Støypet, for example, is expected to shift a boulder with significantly more volume on an igneous rocky shore elsewhere.

Last, the paper by Ruban (2020) [10], which is complimentary to his initial contribution to the SI [3], provides a path forward with examples for future studies of CBDs based on satellite reconnaissance. Such techniques will be critical in the search for additional Neogene localities with mega-clast deposits preserved in coastal settings. Where it focuses on the Middle Miocene climatic optimum and later Pliocene climatic optimum [2], this agenda is sure to prove useful as climatologists make comparisons with the past to gauge the future consequences of global warming related to the erosive power of hurricanes.

**Author Contributions:** The text was prepared by first author (M.E.J.) and approved by the second author (J.L.-V.). All authors have read and agreed to the published version of the manuscript.

**Funding:** This research received no external funding.

**Acknowledgments:** The guest editors for this SI wish to thank the individual authors who participated in the project for their enthusiasm and timely adherence to deadlines.

**Conflicts of Interest:** The authors declare no conflict of interest.

## References

1.  Ogg, J.G.; Ogg, G.; Gradstein, F.M. *The Concise Geologic Time Scale*; Cambridge University Press: Cambridge, UK, 2008; 177p.
2.  Johnson, M.E. Geological Oceanography of the Pliocene Warm Period: A Review with Predictions on the Future of Global Warming. *J. Mar. Sci. Eng.* **2021**, *9*, 1210. [CrossRef]
3.  Ruban, D.M. Coastal Boulder Deposits of the Neogene World: A Synopsis. *J. Mar. Sci. Eng.* **2019**, *7*, 446. [CrossRef]
4.  Guardado-France, R.; Johnson, M.E.; Ledesma-Vázquez, J.; Santa Rosa-del Rio, M.A.; Herrera-Gutiérrez, Á.R. Multiphase Storm Deposits Eroded from Andesite Sea Cliffs on Isla San Luis Gonzaga (Northern Gulf of California, Mexico). *J. Mar. Sci. Eng.* **2020**, *8*, 525. [CrossRef]
5.  Johnson, M.E.; Johnson, E.M.; Guardado-France, R.; Ledesma-Vázquez, J. Holocene Hurricane Deposits Eroded as Coastal Barriers from Andesite Sea Cliffs at Puerto Escondido (Baja California Sur, Mexico). *J. Mar. Sci. Eng.* **2020**, *8*, 75. [CrossRef]
6.  Callahan, G.; Johnson, M.E.; Guardado-France, R.; Ledesma-Vázquez, J. Upper Pleistocene and Holocene Storm Deposits Eroded from the Granodiorite Coast on Isla San Diego (Baja California Sur, Mexico). *J. Mar. Sci. Eng.* **2021**, *9*, 555. [CrossRef]
7.  Ávila, S.P.; Johnson, M.E.; Rebelo, A.C.; Baptista, L.; Melo, C.S. Comparison of Modern and Pleistocene (MIS 5e) Coastal Boulder Deposits from Santa Maria Island (Azores Archipelago, NE Atlantic Ocean). *J. Mar. Sci. Eng.* **2020**, *8*, 386. [CrossRef]

8. Galindo, I.; Johnson, M.E.; Martín-González, E.; Romero, C.; Vegas, J.; Melo, C.S.; Ávila, S.P.; Sánchez, N. Late Pleistocene Boulder Slumps Eroded from a Basalt Shoreline at El Confital Beach on Gran Canaria (Canary Islands, Spain). *J. Mar. Sci. Eng.* **2021**, *9*, 138. [CrossRef]
9. Johnson, M.E. Holocene Boulder Beach Eroded from Chromite and Dunite Sea Cliffs at Støypet on Leka Island (Northern Norway). *J. Mar. Sci. Eng.* **2020**, *8*, 644. [CrossRef]
10. Ruban, D.A. Finding Coastal Megaclast Deposits: A Virtual Perspective. *J. Mar. Sci. Eng.* **2020**, *8*, 164. [CrossRef]

Journal of
*Marine Science and Engineering*

*Article*

# Coastal Boulder Deposits of the Neogene World: A Synopsis

**Dmitry A. Ruban** [1,2,3]

[1]  K.G. Razumovsky Moscow State University of Technologies and Management (the First Cossack University), Zemlyanoy Val Street 73, Moscow 109004, Russia; ruban-d@mail.ru
[2]  Southern Federal University, 23-ja Linija Street 43, Rostov-on-Don 344019, Russia
[3]  Cherepovets State University, Sovetskiy Avenue 10, Cherepovets, Vologda Region 162600, Russia

Received: 31 October 2019; Accepted: 4 December 2019; Published: 5 December 2019

**Abstract:** Modern geoscience research pays significant attention to Quaternary coastal boulder deposits, although the evidence from the earlier geologic periods can be of great importance. The undertaken compilation of the literature permits to indicate 21 articles devoted to such deposits of Neogene age. These are chiefly case studies. Such an insufficiency of investigations may be linked to poor preservation potential of coastal boulder deposits and methodological difficulties. Equal attention has been paid by geoscientists to Miocene and Pliocene deposits. Taking into account the much shorter duration of the Pliocene, an overemphasis of boulders of this age becomes evident. Hypothetically, this can be explained by more favorable conditions for boulder formation, including a larger number of hurricanes due to the Pliocene warming. Geographically, the studies of the Neogene coastal boulder deposits have been undertaken in different parts of the world, but generally in those locations where rocky shores occur nowadays. The relevance of these deposits to storms and tsunamis, rocky shores and deltas, gravity processes, and volcanism has been discussed; however, some other mechanisms of boulder production, transportation, and accumulation (e.g., linked to seismicity and weathering) have been missed.

**Keywords:** bibliography; large clasts; Miocene; Pliocene; rocky shore; storm; tsunami

---

## 1. Introduction

Modern marine sedimentology grows rapidly, and new research directions have strengthened in the past two decades. One of such directions embraces studies of coastal boulders. On the one hand, these studies aim at development of nomenclature of large clasts. Concerning nomenclature, some advances have been made in the works by Blair and McPherson [1], Blott and Pye [2], Bruno and Ruban [3], and Terry and Goff [4]. On the other hand, boulders are regarded as precious evidence of present and ancient rocky shore facies and extreme events (such as storms, tsunamis, hurricanes, typhoons, and cyclones). This evidence has been examined by Autret et al. [5], Bhatt et al. [6], Biolchi et al. [7], Cox et al. [8–10], Dawson [11], Engel et al. [12], Erdmann et al. [13], Hearty and Tormey [14], Herterich et al. [15], Hongo et al. [16], Johnson et al. [17,18], Kennedy et al. [19], Kortekaas and Dawson [20], Lau et al. [21], Olsen et al. [22], Paris et al. [23], Pepe et al. [24], Scheffers et al. [25], Schneider et al. [26], Shah-Hosseini et al. [27], Suanez et al. [28], Terry and Goff [29], Terry et al. [30], Trenhaile [31], Watanabe et al. [32], and Weiss and Sheremet [33]. Most probably, the devastating catastrophes like the 2004 Indian Ocean tsunami [34] and the 2011 Tohoku tsunami [35] have fueled the interest of researchers in coastal sedimentology and, particularly, large clasts [36]. Evidently, investigations of the two noted issues often interconnect.

A significant amount of information about coastal boulders has accumulated, and it appears to be highly important to systematize it for further critical analysis and conceptualization. Such an approach is very common in social sciences [37–39], although geoscientists, unfortunately, often underestimate its potential. Marine sedimentologists need a simple guide permitting orientation in the growing research direction. The objective of the present paper is to offer an overview of the literature on Neogene coastal boulder deposits with some inferences on the current state of research. The importance of such a bibliographical analysis in megaclast research was demonstrated earlier [40]. The peculiarities of this paper are triplicate. First, it presents a synopsis summarizing the already-published data. Second, it focuses on the principal literature sources on the noted subjects, which means articles in international journals accessible via major bibliographical databases and considering large clasts in their title, abstract, and/or keywords. Third, this paper deals with the only Neogene Period, the sedimentary record of which is significantly more representative than that of the earlier periods, but differs from the Quaternary coastal deposits.

## 2. Conceptual Basis

### 2.1. Terminology

The focus of the present paper requires clear definition of several terms, from which two principal terms are "boulder" and "coastal boulder deposit". Evidently, the former indicates a sedimentary particle (clast, grain), and the latter indicates a specific sediment type consisting of (dominated by) such particles.

In the "classical" geological literature, the term "boulder" refers to particles larger than 256 mm (e.g., according to the widely used Udden–Wentworth classification scheme) [3]. However, intensification of studies of large clasts occurring in storm- and tsunami-related deposits raised the question of a more detailed nomenclature of such sedimentary particles. In 1999, Blair and McPherson [1] proposed a nomenclature of large clasts and limited the upper size of boulders to ~4 m (larger clasts are blocks). Different approaches were proposed later [2–4]. Of special interest is the distinction between boulders, mesoboulders, and macroboulders attributing to different categories [4]. Boulders are also opposed to megaclasts (Figure 1). Up to now, there is not broad, international agreement of how large clasts should be termed. As a result of this, it is not a mistake to use the very general term "boulder" for all sedimentary particles larger than 256 mm, except for only some specific studies focusing on the nomenclature development or devoted to a very particular size category of large clasts. In the present paper, this term is considered in such a broad way, and its partial substitutes (like megaclasts) are also taken into account.

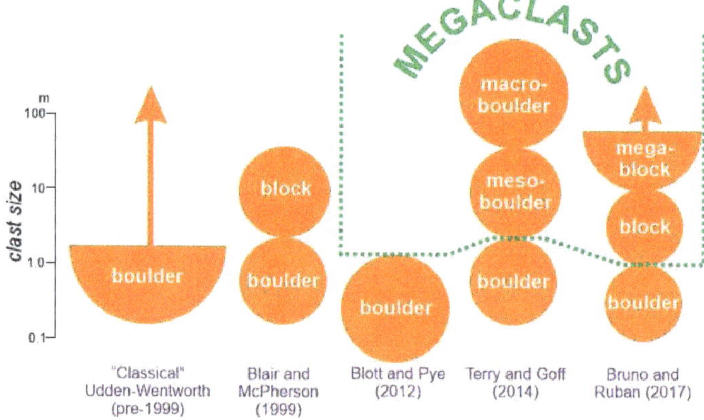


**Figure 1.** Different definition of boulders (see text for references).

The term "coastal boulder deposit" has been used in the works of several authors, although it still requires proper definition. Consideration of the context of its usage in the journal articles permits outlining some characteristics of such a sedimentary formation. These include, particularly, accumulation of boulders distinguished by their large size and/or huge weight [9,10,17,18,25–27], angularity with certain roundness [8], high-topography and inland occurrence [8,9,25,27], high-energy coastlines [5,8,15,17], and relevance to storm and tsunami activity [9,10,18,26,27]. It is notable that the previous works focused more on boulders individually rather than on entire deposits. Even the superficial analysis of the available literature implies that one should distinguish coastal boulder-dominated deposits, i.e., deposits consisting chiefly of boulders (say with their amount of >50%), from coastal boulder-bearing deposits, i.e., deposits dominated by sedimentary particles of lesser size (sand, gravel, cobble, etc.) and bearing a small number or individual boulders. Interestingly, such individual boulders, if too large in size, may look sediment-dominating. The problem seems to be even more complicated in the case of ancient deposits. Large clasts themselves are subject to erosion in the coastal zone with active hydrodynamics, and, thus, large clasts tend to disappear quickly from the geological record. As a result, an ancient coastal boulder-bearing deposit may be legacy of the really existed boulder-dominated deposit. Until these problems are resolved and the nomenclature of large-clast deposits is fixed, it is possible to apply the general term "coastal boulder deposit" broadly, but preferably in those cases when boulders tend to concentrate.

Dewey and Ryan [41] introduced the term "boulderite". Evidently, this can be applied to boulder-dominated deposits. Importantly, the both modern and ancient deposits of this type are called as boulderites [41]. It is the right of the noted authors to use it so, although one may question whether the term "boulderite" can be used for only ancient boulder-dominated deposits, i.e., sedimentary rocks, not recent sediments.

### 2.2. Stratigraphical Framework

The Neogene Period lasted ~20.5 Ma, and it is subdivided into the Miocene and Pliocene Epochs. After strong disputes in the 2000s when the Neogene was extended to the Holocene, a "classical" (almost) scheme has been fixed by the International Commission on Stratigraphy (Table 1), although it is not excluded that the formal definition of the Anthropocene would result in reorganization of the Late Cenozoic stratigraphical nomenclature with subsequent changes in the extent of the Neogene.

**Table 1.** Current version of the Neogene time scale (after International Commission on Stratigraphy [42]).

| Eon | Era | Period | Epoch | Stage | Numerical Age (Ma) of Stage Start |
|---|---|---|---|---|---|
| | | | Quaternary | | 2.580 |
| | | | Pliocene | Piacenzian | 3.600 |
| | | | | Zanclean | 5.333 |
| | | | | Messinian | 7.246 |
| | | | | Tortonian | 11.63 |
| Phanerozoic | Cenozoic | Neogene | Miocene | Serravallian | 13.82 |
| | | | | Langhian | 15.97 |
| | | | | Burdigalian | 20.44 |
| | | | | Aquitanian | 23.03 |
| | | | Paleogene | | 66.00 |

What is necessary to note is the significant disproportion of the Neogene subdivision: the Miocene constitutes ~87% of the period length, and, thus, the Pliocene seems to be too short. This fact should be taken into account when the temporal distribution of any class of geological objects like coastal boulder deposits is analyzed by epochs.

## 3. Bibliographical Synopsis

### 3.1. Research Foci

Although coastal boulder deposits are mentioned in the modern geoscience literature not so rarely, the majority of works deal with the recent and Quaternary boulders. The knowledge of the Neogene sediments of this type remains very restricted. The number of the principal sources does not exceed two dozen. Most probably, this reflects the both low preservation potential of large clasts that themselves are subject of erosion and destruction starting immediately after their deposition and the absence of well-known and broadly accepted techniques for their investigations in the geological record. Nonetheless, the research in the Neogene coastal boulder deposits intensified in the 2010s when up to a half of these principal works were published (Table 2).

**Table 2.** General information about the localities considered in the main articles on Neogene coastal boulder deposits (see also Tables 3–5 for terms, ages, and depositional environments).

| Work | Locality ID | Location and/or Formation | Context of Study |
|---|---|---|---|
| Aguirre and Jimenez, 1997 [43] | 1 | Almeria-Nijar Basin | Palaeobiological: hard-bottom coastal communities |
| Allen et al., 2007 [44] | 2 | Manukau Subgroup | Sedimentological: submarine volcaniclastic deposition |
| Cantalamessa and Di Celma, 2005 [45] | 3 | Mejillones Peninsula | Sedimentological: tsunami backwash deposits |
| Dewey and Ryan, 2017 [41] | 4 | Matheson Formation | Sedimentological: deposition under extreme conditions |
| Edwards et al., 2004 [46] | 5 | Lady Julia Percy Island | Sedimentological and geomorphological: volcanic environment |
| Emhoff et al., 2012 [47] | 6 | Isla Cerralvo, Baja California Sur | Stratigraphical and sedimentological: massive crushed-rhodolith deposit |
| Gutierrez-Mas and Mas, 2013 [48] | 7 | Gulf of Cadiz | Sedimentological: deposition under extreme conditions |
| Hanken et al., 1996 [49] | 8 | Northeast Rhodes | Sedimentological: deposition in coastal graben |
| Hartley et al., 2001 [50] | 9 | Hornitos; La Portada Formation | Sedimentological: tsunamite |
| Hood and Nelson, 2012 [51] | 10 | eastern Taranaki Basin | Sedimentological: carbonate debrites and tectonic control |
| Johnson, 2006 [52] | | global | Sedimentological and palaeobiological: rocky shores and their ecosystems |
| Johnson et al., 2011 [53] | 11 | Madeira Archipelago | Sedimentological and palaeobiological: rhodolith transport |
| Johnson et al., 2012 [54] | 6 | Isla Cerralvo, Baja California Sur | Sedimentological and palaeobiological: rhodolith stranding event |
| Le Roux et al., 2004 [55] | 12 | Coquimbo Formation | Sedimentological: scarp-controlled rocky shoreline |
| Roberts and Brink, 2002 [56] | 13 | Western Cape; Prospect Hill Formation | Stratigraphical: dating of coastal deposits |
| Rodriguez-Tovar et al., 2015 [57] | 14 | Sorbas basin | Palaeobiological: borings in gneiss boulders |
| Shiki and Yamazaki, 1996 [58] | 15 | Chita Peninsula; Morozaki Group | Sedimentological: upper bathyal tsunamites |
| Tachibana and Tsuji, 2011 [59] | 15 | Chita Peninsula; Morozaki Group | Sedimentological: upper bathyal tsunamites |
| Watkins, 1992 [60] | 16 | Salton Trough region; Imperial Formation | Sedimentological and palaeobiological: shallow marine conglomerates and the relevant communities |
| Wesselingh et al., 2013 [61] | 17 | Balgoy; Oosterhout Formation | Palaeobiological: brachiopod-dominated sea-floor assemblage from hardened sandstone boulders |
| Winn and Pousai, 2010 [62] | 18 | Papuan Peninsula; Orubadi and Era Formations | Sedimentological: alluvial-fan and fan-delta deposition |

Interestingly, different researchers use different terminology (Table 3). The majority informs about boulders. In only one case megaclasts are mentioned. Coastal boulder deposits are indicated in five works, although in none of them the term "coastal boulder deposit" is used. These deposits are recognized as boulder beach, boulder conglomerate, or boulderite. Boulder-bearing conglomerate and breccia are also considered, but these should be distinguished from boulder-dominated deposits (see terminological notes above). Finally, a few works employ two or even three terms simultaneously.

**Table 3.** Coastal boulder-related terminology in the main articles on Neogene coastal boulder deposits.

| Work | Basic Terms | | | |
|---|---|---|---|---|
| | Boulder | Coastal Boulder Deposit | Megaclast | Other |
| Aguirre and Jimenez, 1997 [43] | + | | | |
| Allen et al., 2007 [44] | + | | | |
| Cantalamessa and Di Celma, 2005 [45] | + | | | boulder-bearing breccia |
| Dewey and Ryan, 2017 [41] | + | boulderite | + | |
| Edwards et al., 2004 [46] | | boulder beach | | |
| Emhoff et al., 2012 [47] | + | | | |
| Gutierrez-Mas and Mas, 2013 [48] | + | | | |
| Hanken et al., 1996 [49] | | boulder beach | | |
| Hartley et al., 2001 [50] | + | | | |
| Hood and Nelson, 2012 [51] | + | | | |
| Johnson, 2006 [52] | + | | | |
| Johnson et al., 2011 [53] | + | | | |
| Johnson et al., 2012 [54] | + | | | |
| Le Roux et al., 2004 [55] | + | | | |
| Roberts and Brink, 2002 [56] | | boulder beach | | |
| Rodriguez-Tovar et al., 2015 [57] | + | | | |
| Shiki and Yamazaki, 1996 [58] | | | | boulder-bearing conglomerate |
| Tachibana and Tsuji, 2011 [59] | + | | | |
| Watkins, 1992 [60] | + | boulder conglomerate | | |
| Wesselingh et al., 2013 [61] | + | | | |
| Winn and Pousai, 2010 [62] | + | | | |

The majority of the works are case studies focusing on a given location and given stratigraphical intervals. Only two papers of general kind (conceptual) are found (Table 4). The first is the synthetic work of Johnson [52] who overviewed the knowledge of rocky shorelines where boulders often accumulate and the relevant palaeoecosystems. Particularly, he noted that the Neogene deposits of this facies are often linked to ramps, in contrast to the dominance of terrace deposits in the Pleistocene. The second paper of this kind can be judged conceptual only provisionally because this is dealing with the comparison of the examples of the modern and Neogene coastal boulder deposits with a discussion of their storm versus tsunami origin [41]. Importantly, this paper [41] employs the term "boulderite" as equivalent to "boulder-dominated deposit". The other works explore some particular aspects of Neogene coastal boulder deposits, including their relevance to extreme events such as storms and tsunamis, as well as palaeoecological issues.

**Table 4.** Stratigraphical and geographical foci of the main articles on Neogene coastal boulder deposits.

| Work | Conceptual | Miocene | Pliocene | Location |
|---|---|---|---|---|
| Aguirre and Jimenez, 1997 [43] | | | + | Spain |
| Allen et al., 2007 [44] | | + | | New Zealand |
| Cantalamessa and Di Celma, 2005 [45] | | + | | Chile |
| Dewey and Ryan, 2017 [41] | + | + | | New Zealand |
| Edwards et al., 2004 [46] | | + | | Australia (south) |
| Emhoff et al., 2012 [47] | | | + | Mexico |
| Gutierrez-Mas and Mas, 2013 [48] | | | + | Spain |
| Hanken et al., 1996 [49] | | | + | Greece (Rhodes) |
| Hartley et al., 2001 [50] | | | + | Chile |
| Hood and Nelson, 2012 [51] | | + | | New Zealand |
| Johnson, 2006 [52] | + | + | + | World |
| Johnson et al., 2011 [53] | | + | | Portugal (Madeira) |
| Johnson et al., 2012 [54] | | | + | Mexico |
| Le Roux et al., 2004 [55] | | + | + | Chile |
| Roberts and Brink, 2002 [56] | | + | | South Africa |
| Rodriguez-Tovar et al., 2015 [57] | | + | | Spain |
| Shiki and Yamazaki, 1996 [58] | | + | | Japan |
| Tachibana and Tsuji, 2011 [59] | | + | | Japan |
| Watkins, 1992 [60] | | | + | USA (California) |
| Wesselingh et al., 2013 [61] | | | + | Netherlands |
| Winn and Pousai, 2010 [62] | | | + | Papua New Guinea |

It is possible to classify all principal sources on the basis of their stratigraphical, geographical, and genetic foci (Tables 4 and 5). The main observations are as follows. First, Miocene and Pliocene coastal boulder deposits have been generally considered with attention (Table 4). Second, the relevant studies tend to represent different parts of the world (Table 4). Third, the diversity of the discussed mechanisms leading to boulder production, transportation, and accumulation in coastal zone is moderate if not low (Table 5).

**Table 5.** Genetic focus of the main articles on Neogene coastal boulder deposits.

| Work | Rocky Shore | Storm (S), Tsunami (T) | Delta, Fan | Volcanism | Gravity Movement |
|---|---|---|---|---|---|
| Aguirre and Jimenez, 1997 [43] | | | + | | + |
| Allen et al., 2007 [44] | | | | + | |
| Cantalamessa and Di Celma, 2005 [45] | | T | | | + |
| Dewey and Ryan, 2017 [41] | | S, T | | | |
| Edwards et al., 2004 [46] | | | | + | |
| Emhoff et al., 2012 [47] | | | + | | |
| Gutierrez-Mas and Mas, 2013 [48] | | S, T | | | |
| Hanken et al., 1996 [49] | | | not specified | | |
| Hartley et al., 2001 [50] | | T | + | | |
| Hood and Nelson, 2012 [51] | | S | | | + |
| Johnson, 2006 [52] | + | | | | |
| Johnson et al., 2011 [53] | + | S | | + | |
| Johnson et al., 2012 [54] | + | S | + | | |
| Le Roux et al., 2004 [55] | + | | | | + |
| Roberts and Brink, 2002 [56] | | | not specified | | |
| Rodriguez-Tovar et al., 2015 [57] | | | not specified | | |
| Shiki and Yamazaki, 1996 [58] | | T | | | |
| Tachibana and Tsuji, 2011 [59] | | T | | | |
| Watkins, 1992 [60] | + | | | | + |
| Wesselingh et al., 2013 [61] | | S | | | |
| Winn and Pousai, 2010 [62] | | | + | | + |

### 3.2. Further Inferences

The Miocene coastal boulder deposits are considered in 57% of the analyzed works, and those Pliocene are considered in 52% of the works (two articles deal with the both epochs). Apparently, this means equal attention to the both epochs. However, it is necessary to take into account that the Miocene is by ~6.5 times longer than the Pliocene (Table 1). In regard to this fact, it is possible to conclude about significant overemphasis on the Pliocene coastal boulders. Although it cannot be excluded that such a disproportion results from occasional bias in the international research, it can be also hypothesized that the Pliocene environment was more favorable for production and accumulation of boulders in coastal zones of seas and oceans. The evidence of a potentially greater number of hurricanes under the conditions of the Pliocene warming [63–65] makes this hypothesis meaningful. For coastal zone dynamics, sea-level fluctuations seem to be important control of boulder production. Rising sea level accelerates abrasion (especially of sea cliffs) and also leads to growth of shoreline length. For instance, boulders are reported from some areas that were embraced by the sea in the Neogene, but are located inland nowadays, as in the case of the Sorbas Basin in Spain [57]. The global sea level was rather high in the Miocene, but it experienced significant fluctuations that intensified in the Pliocene [66–71]. On the one hand, the relevant instability of the coastal zones could contribute to more boulder formation. On the other hand, the same instability could trigger boulder motion and destruction by waves.

The geographical distribution of the reported Neogene coastal boulder deposits is broad (Figure 2). Despite the rarity of the described locations, the latter occur in all parts of the world (except for Antarctica). It is notable that these deposits have been described chiefly in the same regions where the modern rocky coasts with boulders exist. This is not surprising because of the absence of too striking differences in the position of continents and oceans between the Neogene and the Recent. However, another, complex explanation can be proposed. Sedimentologists and geomorphologists specialized in the studies of coastal boulder deposits often deal with the modern objects. If so, it is evident that they are able to detect ancient deposits of this kind in the same geographical loci. Nonetheless, it is evident that the knowledge of Quaternary coastal boulders is much wider. For instance, these have been reported from many localities of the Mediterranean, including (but not limited to) Istria [7], Sicily [24], northern Egypt [27], Malta [72], Ibiza [73], Crete [74], Lesvos [75], southern France [76], and Apulia [77]. Better to say, boulders and their accumulations are found on the majority of coasts of the Mediterranean Sea. In contrast, Neogene large clasts are reported from very few localities of the same basin. Most probably, this reflects the both sedimentological research bias and low preservation potential of boulders.

An interesting inference is linked to the origin of the Neogene coastal boulder deposits. Many previous studies focused on their relevance to storms and tsunamis as the leading boulder production, transportation, and accumulation forces, as well as on gravity processes linked to downslope movement with consequent cliff retreat (Figure 3). The main depositional environments analyzed in the course of the coastal boulder research are rocky shores, deltas, and areas of volcanism (Figure 3). On the one hand, it is clear that chiefly extreme events like storms, tsunamis, and volcanic eruptions are able to provide the energy necessary to produce and to move large clasts. On the other hand, it seems to be questionable if some other forces were responsible. For instance, seismicity would cause giant cliff collapse or heterogeneity of exposed substrate would lead to its differential erosion. Finally, what about the possible role of wind erosion in coastal zones? Undoubtedly, identification and correct interpretation of such phenomena even in geological records as young as that of the Neogene is highly challenging and requires very creative analysis. Examples of the latter can be found in the works deciphering the origin of boulders from the Miocene upper bathyal deposits of the Chita Peninsula (Japan) [58,59] ensures the possibility of such state-of-the-art investigations. Anyway, coastal boulders, especially those measured by meters are highly specific and uncommon geological objects, and their analysis should be undertaken in regard to individual peculiarities of each given locality.

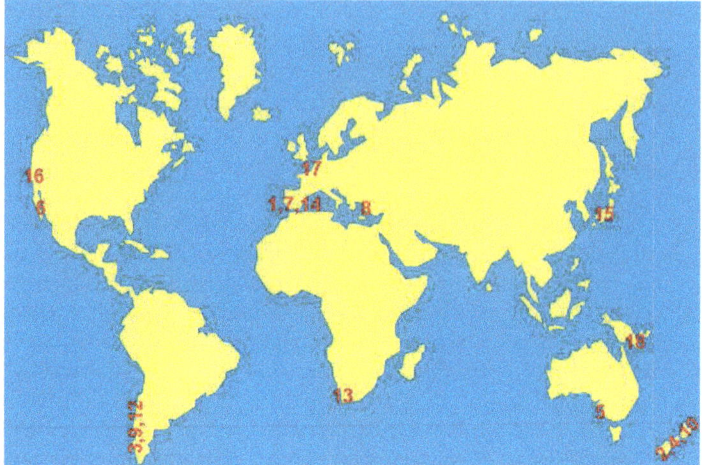

**Figure 2.** Geographical focus of the studies of Neogene coastal boulder deposits (based on Table 4). See Table 2 for locality IDs.

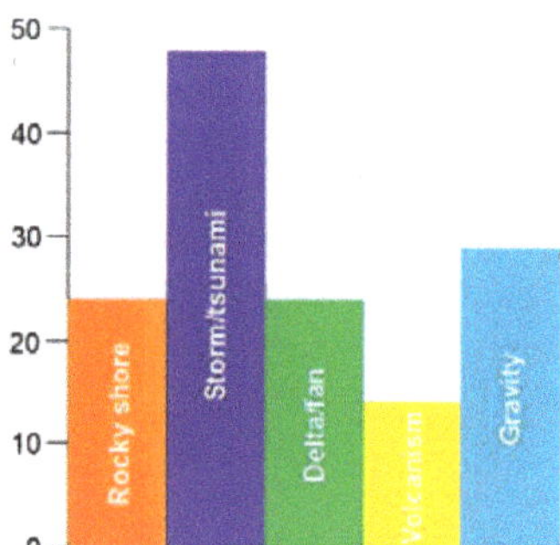


**Figure 3.** Genetic focus of the studies of Neogene coastal boulder deposits (based on Table 5).

## 4. Conclusions

The bibliographical synopsis of the knowledge of Neogene coastal boulder deposits implies that the relevant research has been weak. Nonetheless, this research has generated significant evidence of these deposits. The main findings of the present analysis are as follows.

(1) Case studies of the Neogene coastal boulder deposits prevail over conceptual works.

(2) Attention has been paid to the both epochs of the Neogene (although with overemphasis on the Pliocene), to many parts of the world, and to the really principal mechanisms of boulder production, transportation, and accumulation (first of all, to extreme events).

(3) The stratigraphical, geographical, and genetic foci of the research demonstrate certain biases that can be explained, particularly, by peculiarities of the geological record.

Generally, this means that although the Neogene coastal boulder deposits are highly specific and rather uncommon geological objects, the latter have been studied more or less adequately to make further interpretations of their relevance to the dynamics of the Neogene world.

**Funding:** This research received no external funding.

**Acknowledgments:** The author gratefully thanks M.E. Johnson (USE) for his kind invitation to contribute to this special issue and various support, as well as the reviewers for their helpful suggestions.

**Conflicts of Interest:** The author declares no conflict of interest.

## References


1. Blair, T.C.; McPherson, J.G. Grain-size and textural classification of coarse sedimentary particles. *J. Sediment. Res.* **1999**, *69*, 6–19. [CrossRef]
2. Blott, S.J.; Pye, K. Particle size scales and classification of sediment types based on particle size distributions: Review and recommended procedures. *Sedimentology* **2012**, *59*, 2071–2096. [CrossRef]
3. Bruno, D.E.; Ruban, D.A. Something more than boulders: A geological comment on the nomenclature of megaclasts on extraterrestrial bodies. *Planet. Space Sci.* **2017**, *135*, 37–42. [CrossRef]
4. Terry, J.P.; Goff, J. Megaclasts: Proposed revised nomenclature at the coarse end of the Udden-Wentworth grain-size scale for sedimentary particles. *J. Sediment. Res.* **2014**, *84*, 192–197. [CrossRef]
5. Autret, R.; Dodet, G.; Suanez, S.; Roudaut, G.; Fichaut, B. Long–term variability of supratidal coastal boulder activation in Brittany (France). *Geomorphology* **2018**, *304*, 184–200. [CrossRef]
6. Bhatt, N.; Murari, M.K.; Ukey, V.; Prizomwala, S.P.; Singhvi, A.K. Geological evidences of extreme waves along the Gujarat coast of western India. *Nat. Hazards* **2016**, *84*, 1685–1704. [CrossRef]
7. Biolchi, S.; Furlani, S.; Devoto, S.; Scicchitano, G.; Korbar, T.; Vilibic, I.; Sepic, J. The origin and dynamics of coastal boulders in a semi-enclosed shallow basin: A northern Adriatic case study. *Mar. Geol.* **2019**, *411*, 62–77. [CrossRef]
8. Cox, R.; Lopes, W.A.; Jahn, K.L. Quantitative roundness analysis of coastal boulder deposits. *Mar. Geol.* **2018**, *396*, 114–141. [CrossRef]
9. Cox, R.; Jahn, K.L.; Watkins, O.G.; Cox, P. Extraordinary boulder transport by storm waves (west of Ireland, winter 2013–2014), and criteria for analysing coastal boulder deposits. *Earth Sci. Rev.* **2018**, *177*, 623–636. [CrossRef]
10. Cox, R.; O'Boyle, L.; Cytrynbaum, J. Imbricated Coastal Boulder Deposits are Formed by Storm Waves, and Can Preserve a Long-Term Storminess Record. *Sci. Rep.* **2019**, *9*, 10784. [CrossRef]
11. Dawson, A. The geological significance of tsunamis. *Z. Fur Geomorphol. Suppl.* **1996**, *102*, 199–210.
12. Engel, M.; Oetjen, J.; May, S.M.; Bruckner, H. Tsunami deposits of the Caribbean – Towards an improved coastal hazard assessment. *Earth Sci. Rev.* **2016**, *163*, 260–296. [CrossRef]
13. Erdmann, W.; Scheffers, A.M.; Kelletat, D.H. Holocene Coastal Sedimentation in a Rocky Environment: Geomorphological Evidence from the Aran Islands and Galway Bay (Western Ireland). *J. Coast. Res.* **2018**, *34*, 772–792. [CrossRef]
14. Hearty, P.J.; Tormey, B.R. Sea-level change and superstorms; geologic evidence from the last interglacial (MIS 5e) in the Bahamas and Bermuda offers ominous prospects for a warming Earth. *Mar. Geol.* **2017**, *390*, 347–365. [CrossRef]
15. Herterich, J.G.; Cox, R.; Dias, F. How does wave impact generate large boulders? Modelling hydraulic fracture of cliffs and shore platforms. *Mar. Geol.* **2018**, *399*, 34–46. [CrossRef]
16. Hongo, C.; Kurihara, H.; Golbuu, Y. Coral boulders on Melekeok reef in the Palau Islands: An indicator of wave activity associated with tropical cyclones. *Mar. Geol.* **2018**, *399*, 14–22. [CrossRef]
17. Johnson, M.E.; Ledesma-Vazquez, J.; Guardado-France, R. Coastal Geomorphology of a Holocene Hurricane Deposits on a Pleistocene Marine Terrace from Isla Carmen (Baja California Sur, Mexico). *J. Mar. Sci. Eng.* **2018**, *6*, 108. [CrossRef]
18. Johnson, M.E.; Guardado-France, R.; Johnson, E.M.; Ledesma-Vazquez, J. Geomorphology of a Holocene Hurricane Deposit Eroded from Rhyolite Sea Cliffs on Ensenada Almeja (Baja California Sur, Mexico). *J. Mar. Sci. Eng.* **2019**, *7*, 193. [CrossRef]
19. Kennedy, D.M.; Woods, J.L.D.; Naylor, L.A.; Hansom, J.D.; Rosser, N.J. Intertidal boulder-based wave hindcasting can underestimate wave size: Evidence from Yorkshire, UK. *Mar. Geol.* **2019**, *411*, 98–106. [CrossRef]
20. Kortekaas, S.; Dawson, A.G. Distinguishing tsunami and storm deposits: An example from Martinhal, SW Portugal. *Sediment. Geol.* **2007**, *200*, 208–221. [CrossRef]
21. Lau, A.Y.A.; Terry, J.P.; Ziegler, A.; Pratap, A.; Harris, D. Boulder emplacement and remobilisation by cyclone and submarine landslide tsunami waves near Suva City, Fiji. *Sediment. Geol.* **2018**, *364*, 242–257. [CrossRef]
22. Olsen, M.J.; Johnstone, E.; Driscoll, N.; Kuester, F.; Ashford, S.A. Fate and transport of seacliff failure sediment in southern California. *J. Coast. Res.* **2016**, *76*, 185–199. [CrossRef]

23. Paris, R.; Naylor, L.A.; Stephenson, W.J. Boulders as a signature of storms on rock coasts. *Mar. Geol.* **2011**, *283*, 1–11. [CrossRef]
24. Pepe, F.; Corradino, M.; Parrino, N.; Besio, G.; Presti, V.L.; Renda, P.; Calcagnile, L.; Quarta, G.; Sulli, A.; Antonioli, F. Boulder coastal deposits at Favignana Island rocky coast (Sicily, Italy): Litho-structural and hydrodynamic control. *Geomorphology* **2018**, *303*, 191–209. [CrossRef]
25. Scheffers, A.; Kelletat, D.; Haslett, S.; Scheffers, S.; Browne, T. Coastal boulder deposits in Galway Bay and the Aran Islands, western Ireland. *Z. Fur Geomorphol.* **2010**, *54*, 247–279. [CrossRef]
26. Schneider, B.; Hoffmann, G.; Falkenroth, M.; Grade, J. Tsunami and storm sediments in Oman: Characterizing extreme wave deposits using terrestrial laser scanning. *J. Coast. Conserv.* **2019**, *23*, 801–815. [CrossRef]
27. Shah-Hosseini, M.; Saleem, A.; Mahmoud, A.-M.A.; Morhange, C. Coastal boulder deposits attesting to large wave impacts on the Mediterranean coast of Egypt. *Nat. Hazards* **2016**, *83*, 849–865. [CrossRef]
28. Suanez, S.; Fichaut, B.; Magne, R. Cliff-top storm deposits on Banneg Island, Brittany, France: Effects of giant waves in the Eastern Atlantic Ocean. *Sediment. Geol.* **2009**, *220*, 12–28. [CrossRef]
29. Terry, J.P.; Goff, J. Strongly aligned coastal boulders on Ko Larn island (Thailand): A proxy for past typhoon-driven high-energy wave events in the Bay of Bangkok. *Geogr. Res.* **2019**, *57*, 344–358. [CrossRef]
30. Terry, J.P.; Goff, J.; Jankaew, K. Major typhoon phases in the upper Gulf of Thailand over the last 1.5 millennia, determined from coastal deposits on rock islands. *Quat. Int..*
31. Trenhaile, A. Rocky coasts–Their role as depositional environments. *Earth Sci. Rev.* **2016**, *159*, 1–13. [CrossRef]
32. Watanabe, M.; Goto, K.; Imamura, F.; Hongo, C. Numerical identification of tsunami boulders and estimation of local tsunami size at Ibaruma reef of Ishigaki Island, Japan. *Isl. Arc* **2016**, *25*, 316–332. [CrossRef]
33. Weiss, R.; Sheremet, A. Toward a new paradigm for boulder dislodgement during storms. *Geochem. Geophys. Geosyst.* **2017**, *18*, 2717–2726. [CrossRef]
34. Lay, T.; Kanamori, H.; Ammon, C.J.; Nettles, M.; Ward, S.N.; Aster, R.C.; Beck, S.L.; Bilek, S.L.; Brudzinski, M.R.; Butler, R.; et al. The great Sumatra-Andaman earthquake of 26 December 2004. *Science* **2005**, *308*, 1127–1133. [CrossRef] [PubMed]
35. Simons, M.; Minson, S.E.; Sladen, A.; Ortega, F.; Jiang, J.; Owen, S.E.; Meng, L.; Ampuero, J.-P.; Wei, S.; Chu, R.; et al. The 2011 magnitude 9.0 Tohoku-Oki earthquake: Mosaicking the megathrust from seconds to centuries. *Science* **2011**, *332*, 1421–1425. [CrossRef] [PubMed]
36. Ruban, D.A. Research in tsunami-related sedimentology during 2001–2010: Can a single natural disaster re-shape the science? *GeoActa* **2011**, *10*, 79–85.
37. Fernandez, K.V. Critically reviewing literature: A tutorial for new researchers. *Australas. Mark. J.* **2019**, *27*, 187–196. [CrossRef]
38. Kumar, P.; Sharma, A.; Salo, J. A bibliometric analysis of extended key account management literature. *Ind. Mark. Manag.* **2019**, *82*, 276–292. [CrossRef]
39. Snyder, H. Literature review as a research methodology: An overview and guidelines. *J. Bus. Res.* **2019**, *104*, 333–339. [CrossRef]
40. Ruban, D.A.; Ponedelnik, A.A.; Yashalova, N.N. Megaclasts: Term Use and Relevant Biases. *Geosciences* **2019**, *9*, 14. [CrossRef]
41. Dewey, J.F.; Ryan, P.D. Storm, rogue wave, or tsunami origin for megaclast deposits in Western Ireland and North Island, New Zealand? *Proc. Natl. Acad. Sci. USA* **2017**, *114*, E10639–E10647. [CrossRef]
42. International Commission on Stratigraphy. International Chronostratigraphic Chart 2019. Available online: Stratigraphy.org (accessed on 18 October 2019).
43. Aguirre, J.; Jimenez, A.P. Census assemblages in hard-bottom coastal communities: A case study from the Plio-Pleistocene Mediterranean. *Palaios* **1997**, *12*, 598–608. [CrossRef]
44. Allen, S.R.; Hayward, B.W.; Mathews, E. A facies model for a submarine volcaniclastic apron: The Miocene Manukau Subgroup, New Zealand. *Bull. Geol. Soc. Am.* **2007**, *119*, 725–742. [CrossRef]
45. Cantalamessa, G.; Di Celma, C. Sedimentary features of tsunami backwash deposits in a shallow marine Miocene setting, Mejillones Peninsula, northern Chile. *Sediment. Geol.* **2005**, *178*, 259–273. [CrossRef]
46. Edwards, J.; Cayley, R.A.; Joyce, E.B. Geology and geomorphology of the Lady Julia Percy Island volcano, a Late Miocene submarine and subaerial volcano off the coast of Victoria, Australia. *Proc. R. Soc. Vic.* **2004**, *116*, 15–35.

47. Emhoff, K.F.; Johnson, M.E.; Backus, D.H.; Ledesma-Vazquez, J. Pliocene stratigraphy at paredones blancos: Significance of a massive crushed-rhodolith deposit on Isla Cerralvo, baja California sur (Mexico). *J. Coast. Res.* **2012**, *28*, 234–243. [CrossRef]

48. Gutierrez-Mas, J.M.; Mas, R. Record of very high energy events in Plio-Pleistocene marine deposits of the Gulf of Cadiz (SW Spain): Facies and processes. *Facies* **2013**, *59*, 679–701. [CrossRef]

49. Hanken, N.-M.; Bromley, R.G.; Miller, J. Plio-Pleistocene sedimentation in coastal grabens, north-east Rhodes, Greece. *Geol. J.* **1996**, *31*, 393–418. [CrossRef]

50. Hartley, A.; Howell, J.; Mather, A.E.; Chong, G. A possible Plio-Pleistocene tsunami deposit, Hornitos, Northern Chile. *Rev. Geol. Chile* **2001**, *28*, 117–125. [CrossRef]

51. Hood, S.D.; Nelson, C.S. Temperate carbonate debrites and short-lived earliest Miocene yo-yo tectonics, eastern Taranaki Basin margin, New Zealand. *Sediment. Geol.* **2012**, *247–248*, 58–70. [CrossRef]

52. Johnson, M.E. Uniformitarianism as a guide to rocky-shore ecosystems in the geological record. *Can. J. Earth Sci.* **2006**, *43*, 1119–1147. [CrossRef]

53. Johnson, M.E.; da Silva, C.M.; Santos, A.; Baarli, B.G.; Cachao, M.; Mayoral, E.J.; Rebelo, A.C.; Ledesma-Vazquez, J. Rhodolith transport and immobilization on a volcanically active rocky shore: Middle Miocene at Cabeco das Laranjas on Ilheu de Cima (Madeira Archipelago, Portugal). *Palaeogeogr. Palaeoclimatol. Palaeoecol.* **2011**, *300*, 113–127. [CrossRef]

54. Johnson, M.E.; Perez, D.M.; Baarli, B.G. Rhodolith stranding event on a Pliocene rocky shore from Isla Cerralvo in the lower Gulf of California (Mexico). *J. Coast. Res.* **2012**, *28*, 225–233. [CrossRef]

55. Le Roux, J.P.; Gomez, C.; Fenner, J.; Middleton, H. Sedimentological processes in a scarp-controlled rocky shoreline to upper continental slope environment, as revealed by unusual sedimentary features in the Neogene Coquimbo Formation, north-central Chile. *Sediment. Geol.* **2004**, *165*, 67–92. [CrossRef]

56. Roberts, D.L.; Brink, J.S. Dating and correlation of Neogene coastal deposits in the Western Cape (South Africa): Implications for neotectonism. *S. Afr. J. Geol.* **2002**, *105*, 337–352. [CrossRef]

57. Rodriguez-Tovar, F.J.; Uchman, A.; Puga-Bernabeu, A. Borings in gneiss boulders in the Miocene (Upper Tortonian) of the Sorbas basin, SE Spain. *Geol. Mag.* **2015**, *152*, 287–297. [CrossRef]

58. Shiki, T.; Yamazaki, T. Tsunami-induced conglomerates in Miocene upper bathyal deposits, Chita Peninsula, central Japan. *Sediment. Geol.* **1996**, *104*, 175–188. [CrossRef]

59. Tachibana, T.; Tsuji, Y. Geological and hydrodynamical examination of the bathyal tsunamigenic origin of miocene conglomerates in Chita peninsula, Central Japan. *Pure Appl. Geophys.* **2011**, *168*, 997–1014. [CrossRef]

60. Watkins, R. Sedimentology and paleoecology of Pliocene shallow marine conglomerates, Salton Trough region, California. *Palaeogeogr. Palaeoclimatol. Palaeoecol.* **1992**, *95*, 319–333. [CrossRef]

61. Wesselingh, F.P.; Peters, W.J.M.; Munsterman, D.K. A brachiopod-dominated sea-floor assemblage from the Late Pliocene of the eastern Netherlands. *Neth. J. Geosci.* **2013**, *92*, 171–176. [CrossRef]

62. Winn, R.D., Jr.; Pousai, P. Synorogenic alluvial-fan—Fan-delta deposition in the Papuan foreland basin: Plio-Pleistocene Era Formation, Papua New Guinea. *Aust. J. Earth Sci.* **2010**, *57*, 507–523. [CrossRef]

63. Fedorov, A.V.; Brierley, C.M.; Emanuel, K. Tropical cyclones and permanent El Niño in the early Pliocene epoch. *Nature* **2010**, *463*, 1066–1070. [CrossRef] [PubMed]

64. Johnson, M.E.; Uchman, A.; Costa, P.J.M.; Ramalho, R.S.; Ávila, S.P. Intense hurricane transports sand onshore: Example from the Pliocene Malbusca section on Santa Maria Island (Azores, Portugal). *Mar. Geol.* **2017**, *385*, 244–249. [CrossRef]

65. Yan, Q.; Wei, T.; Korty, R.L.; Kossin, J.P.; Zhang, Z.; Wang, H. Enhanced intensity of global tropical cyclones during the mid-Pliocene warm period. *Proc. Natl. Acad. Sci. USA* **2016**, *113*, 12963–12967. [CrossRef] [PubMed]

66. Betzler, C.; Eberli, G.P.; Lüdmann, T.; Reolid, J.; Kroon, D.; Reijmer, J.J.G.; Swart, P.K.; Wright, J.; Young, J.R.; Alvarez-Zarikian, C.; et al. Refinement of Miocene sea level and monsoon events from the sedimentary archive of the Maldives (Indian Ocean). *Prog. Earth Planet. Sci.* **2018**, *5*, 5. [CrossRef]

67. Dumitru, O.A.; Austermann, J.; Polyak, V.J.; Fornós, J.J.; Asmerom, Y.; Ginés, J.; Ginés, A.; Onac, B.P. Constraints on global mean sea level during Pliocene warmth. *Nature* **2019**, *574*, 233–236. [CrossRef]

68. Grant, G.R.; Naish, T.R.; Dunbar, G.B.; Stocchi, P.; Kominz, M.A.; Kamp, P.J.J.; Tapia, C.A.; McKay, R.M.; Levy, R.H.; Patterson, M.O. The amplitude and origin of sea-level variability during the Pliocene epoch. *Nature* **2019**, *574*, 237–241. [CrossRef]

69. Kominz, M.A.; Browning, J.W.; Miller, K.G.; Sugarman, P.J.; Mizintseva, S.; Scotese, C.R. Late Cretaceous to Miocene sea-level estimates from the New Jersey and Delaware coastal plain coreholes: An error analysis. *Basin Res.* **2008**, *20*, 211–226. [CrossRef]

70. Raymo, M.E.; Mitrovica, J.X.; O'Leary, M.J.; Deconto, R.M.; Hearty, P.J. Departures from eustasy in Pliocene sea-level records. *Nat. Geosci.* **2011**, *4*, 328–332. [CrossRef]

71. Raymo, M.E.; Kozdon, R.; Evans, D.; Lisiecki, L.; Ford, H.L. The accuracy of mid-Pliocene d18O-based ice volume and sea level reconstructions. *Earth Sci. Rev.* **2018**, *177*, 291–302. [CrossRef]

72. Causon Deguara, J.; Scerri, S. Ras il-Gebel: An extreme wave-generated bouldered coast at Xghajra (Malta). *World Geomorphol. Landsc.* **2019**, 229–243.

73. Roig-Munar, F.X.; Rodríguez-Perea, A.; Martín-Prieto, J.A.; Gelabert, B.; Vilaplana, J.M. Tsunami boulders on the rocky coasts of Ibiza and Formentera (Balearic Islands). *J. Mar. Sci. Eng.* **2019**, *7*, 327. [CrossRef]

74. Scheffers, A.; Scheffers, S. Tsunami deposits on the coastline of west Crete (Greece). *Earth Planet. Sci. Lett.* **2007**, *259*, 613–624. [CrossRef]

75. Vacchi, M.; Rovere, A.; Zouros, N.; Firpo, M. Assessing enigmatic boulder deposits in NE Aegean Sea: Importance of historical sources as tool to support hydrodynamic equations. *Nat. Hazards Earth Syst. Sci.* **2012**, *12*, 1109–1118. [CrossRef]

76. Piscitelli, A.; Milella, M.; Hippolyte, J.-C.; Shah-Hosseini, M.; Morhange, C.; Mastronuzzi, G. Numerical approach to the study of coastal boulders: The case of Martigues, Marseille, France. *Quat. Int.* **2017**, *439*, 52–64. [CrossRef]

77. Mastronuzzi, G.; Sansò, P. Boulders transport by catastrophic waves along the Ionian coast of Apulia (southern Italy). *Mar. Geol.* **2000**, *70*, 93–103. [CrossRef]

Journal of
*Marine Science and Engineering*

*Article*

# Multiphase Storm Deposits Eroded from Andesite Sea Cliffs on Isla San Luis Gonzaga (Northern Gulf of California, Mexico)

Rigoberto Guardado-France [1], Markes E. Johnson [2,*], Jorge Ledesma-Vázquez [1], Miguel A. Santa Rosa-del Rio [1] and Ángel R. Herrera-Gutiérrez [1]

[1]  Facultad de Ciencias Marinas, Universidad Autónoma de Baja California, Ensenada 22800, Mexico; rigoberto@uabc.edu.mx (R.G.-F.); ledesma@uabc.edu.mx (J.L.-V.); msanta@uabc.edu.mx (M.A.S.R.-d.R.); herrera.angel@uabc.edu.mx (Á.R.H.-G.)
[2]  Department of Geosciences, Williams College, Williamstown, MA 01267, USA
*  Correspondence: mjohnson@williams.edu; Tel.: +14-132-329

Received: 25 May 2020; Accepted: 15 July 2020; Published: 16 July 2020

**Abstract:** The 450-m long spit that extends westward from the northwest corner of Isla San Luis Gonzaga is one of the largest and most complex constructions of unconsolidated cobbles and boulders found anywhere in Mexico's Gulf of California. The material source derives from episodic but intense storm erosion along the island's andesitic cliff face with steep northern exposures. A well-defined marine terrace from the late Pleistocene cuts across the same corner of the island and provides a marker for the subsequent development of the spit that post-dates tectonic-eustatic adjustments. A total of 660 individual andesite clasts from seven transects across the spit were measured for analyses of change in shape and size. These data are pertinent to the application of mathematical formulas elaborated after Nott (2003) and subsequent refinements to estimate individual wave heights necessary for lift from parent sea cliffs and subsequent traction. Although the ratio of boulders to clasts diminishes from the proximal to distal end of the structure, relatively large boulders populate all transects and the average wave height required for the release of joint-bound blocks at the rocky shore amounts to 5 m. Based on the region's historical record of hurricanes, such storms tend to decrease in intensity as they migrate northward through the Gulf of California's 1100-km length. However, the size and complexity of the San Luis Gonzaga spit suggests that a multitude of extreme storm events impacted the island in the upper gulf area through the Holocene time, yielding a possible average growth rate between 7 and 8 m/century over the last 10,000 years. In anticipation of future storms, a system to track the movement of sample boulders should be emplaced on the San Luis Gonzaga spit and similar localities with major coastal boulder deposits.

**Keywords:** coastal boulder deposits; storm waves; hydrodynamic equations; Holocene; western North America

---

## 1. Introduction

The Gulf of California is a narrow, semi-enclosed sea that extends from its opening with the Pacific Ocean for more than 1100 km to the northwest between the Baja California peninsula and the mainland of western Mexico. As many as eight hurricanes form each year between May and October over ocean waters that attain a temperature of 27 °C or higher off the Mexican mainland near a latitude of N 15° [1]. Based on several decades of such data, 50% of such storms turn harmlessly westward into the open Pacific Ocean as they shift northward. Only a few track northeast into the Gulf of California but those that do, such as the September 2014 Hurricane Odile, are capable of causing extensive damage to infrastructure on the peninsula [2]. Odile struck the southern tip of the peninsula as a Category

4 hurricane with sustained winds reaching 215 km/h but diminished to a Category 3 event 24 h later as it tracked into the lower Gulf of California. By the time it reached the upper part of the gulf and crossed into mainland Mexico, the disturbance was reduced to a tropical storm. A detailed analysis of that storm by Gross and Mager (2020) applied mathematical models to reconstruct the impact of known meteorological conditions based on wind speed and wind direction to changes in wave height and the degree to which the water column was agitated as the storm progressed through the gulf's entire length [3]. The study's stated objective was to present a worst-case scenario on the impact of damage to tidal-energy devices that might be employed in the upper Gulf of California. Installations of this kind have yet to be built in the region, which registers tidal ranges on the order of 12 m [4]. In theory, the mechanisms engineered to harness energy from tidal exchange are not as susceptible to wind damage as they are to extreme waves. Beyond its stated purpose [3], the contribution by Gross and Mager (2020) provides the most thorough longitudinal treatment of changing physical parameters related to a major storm event in the Gulf of California.

Infrequent as they may appear on a human time frame, extreme storm events wield a measurable and persistent impact on coastal geomorphology over the long term as registered in deposits of various kinds around the world. Studies on Holocene storm chronology are focused mostly on accumulations preserved in coastal marshes, lagoons, and beach ridges [5]. Less attention has been devoted to deposits that result from the erosional retreat of sea cliffs by recurrent storm events [6–8]. On a regional basis limited to the lower Gulf of California, rocky-shore studies have focused on the Holocene development of such features where the erosion of limestone shores and volcanic sea cliffs composed of rhyolite and andesite resulted in extensive coastal boulder deposits (CBDs) and related coastal barriers [9–11]. Andesite is the most widespread rock type exposed in sea cliffs along the western Gulf of California, accounting for nearly 25% of all shoreline features including beaches and mud flats [12]. Andesite rocky coasts are under-represented compared to granite shores in the upper Gulf of California, but still common.

The goal of this study is to expand on the relationship between coastal erosion of andesite sea cliffs and the development of a massive coastal barrier deposit formed by andesite cobbles and boulders on Isla San Luis Gonzaga in the upper Gulf of California. The methods for analysis of eroded clast shapes and sizes together with estimates on the wave heights necessary for their primary generation follow those in previous contributions [9–11]. The choice of the Gonzaga study site was influenced by the prospect of superior control over the scale of sequential changes in topographic layout. It is expected that Holocene CBDs with a time range through thousands of year duration will offer better insight regarding the intensity of episodic storm events in regions like the Gulf of California otherwise perceived to suffer rare events. Civil engineers involved with planning for infrastructure ranging from artificial harbor facilities and breakwaters to potential power linkages with tidal-energy mechanisms need to be aware of such physical settings with a deep background in coastal geomorphology.

## 2. Geographical and Geological Setting

Located in the upper Gulf of California, the study site within Bahía San Luis Gonzaga is midway between the towns of San Felipe and Bahía de Los Angeles (Figure 1a). San Luis Gonzaga constitutes the area's largest bay (Figure 1b), covering an area of about 36 km$^2$. Isla San Luis Gonzaga sits at the northwest side of the bay, approximately 1.5 km$^2$ in area and rising 140 m above sea level (Figure 1c). Detailed geological mapping of the island and the surrounding region confirms that the local bedrock is formed entirely of andesite flows [13]. The focus of this study is a 450-m long spit formed exclusively of andesite cobbles and boulders that extends westward from the northwest corner of the island. Tracing the phased temporal development of the spit ranks as the project's primary goal, which entails advantages in scale and layout compared to earlier studies of CBDs in the lower Gulf of California [9–11].

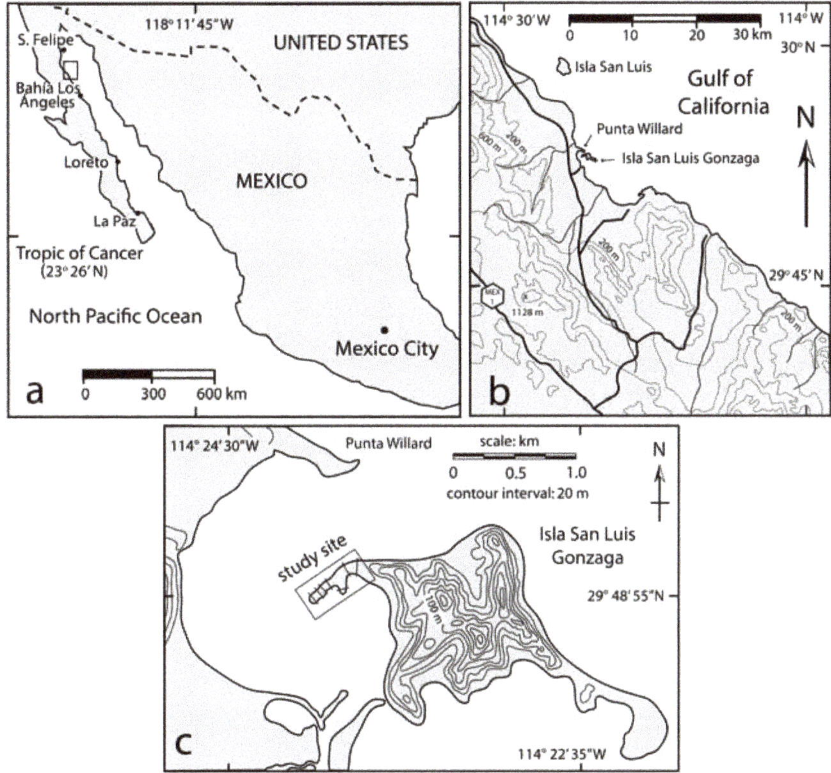

**Figure 1.** Locality maps showing Mexico's Baja California peninsula and Gulf of California; (**a**) Mexico and border area with the United States denoting key towns with inset box marking the study area between San Felipe and Bahía de Los Angeles; (**b**) Region around Bahía San Luis Gonzaga showing Isla San Luis Gonzaga in the northwest part of the bay; (**c**) Topographic map of Isla San Luis Gonzaga with the study site marked (box) within which the study transects are indicated.

## 3. Materials and Methods

### 3.1. Data Collection

Isla San Luis Gonzaga was visited in June 2019, when the original data for this study were collected from unconsolidated andesite clasts forming the long spit attached to the island accessible from Punta Willard (Figure 1c). Cobble- and boulder-size clasts encountered on tape lines through seven transects were measured manually in three dimensions perpendicular to one another in each clast (long, intermediate, and short). All transects were laid out to cross the spit at different locations, always at right angles to the defining shore with orientations recorded by compass. Continuous tracking of elevation with respect to sea level was monitored across each transect in order to construct topographic relief profiles (see Section 3.2 for more details). Differentiated from cobbles, the base definition for a boulder adapted in this exercise is that of Wentworth (1922) for an erosional clast equal or greater than 256 mm in diameter [14]. No upper limit for this category is defined in the geological literature. Triangular plots were employed to show variations in clast shape, following the design of Sneed and Folk (1958) for river pebbles [15]. Comparative data on maximum cobble and boulder dimensions were fitted to bar graphs to show variations in composition from one transect to the next. Multiple samples of andesite were collected from the island's rocky-shore zone for laboratory analysis to determine specific gravity.

### 3.2. Aerial Photography and Applications for Topography

Use of a DJI Inspire 2 Drone™ (DJI, Nanshan District, Shenzhen, China) was employed to generate the Geographic Information System (GIS) platform, photogrammetry, and digital elevation model (DEM) in this study following standard protocols [16]. Eight tarps, each covering 1 m$^2$, were printed with a highly visible pattern that could clearly be detected from the sky. The tarps were laid out across the rocky bar and georeferenced using a handheld Geographic Positioning System (GPS). Thereafter, a flight plan was designed and uploaded to the drone with a limited flight duration between 23 and 27 min that was ample for completion while providing a stable platform. The flight plan was designed to calculate the number of images the drone needed to take based on altitude and desired image overlay (70–80%). Each image captured by the drone was automatically georeferenced and transferred to photogrammetry software, where steps were followed to join the images into a single mosaic. The first step is image alignment, wherein the software places and aligns the images taken by the drone based on the GPS data from the flight plan, as well as the control points from the tarps and GPS data collected on sight. The next step entailed object identification so that tie points could be generated to stitch the images together. A sparse cloud was next generated from a series of points using the overlaid images, points in common, and drone flight data to calculate the elevation of each point. With these data from the dense cloud, a mesh is created as a series of triangles that joins the points from the dense cloud, and the resulting layer is a continuous surface on which the original images can be "draped over". Based on the generated data in the previous step, a DEM can be generated using the kriging interpolation method. For example, the same strategy has a successful application for high accuracy surveying of beach-sand topography [17].

Using GIS software by Agisoft Metashape, the DEM was applied to determine the slope and direction of the slope traversing the spit. To determine slope the software calculates the angle on the incline based on the elevation of each pixel and its relation to the adjoining pixels. With the slope layer generated and geographic location of the layer the software determines the downslope direction for each cell within the DEM. The resulting layer indicates the main slope directions of the feature. To obtain the elevation of the associated marine terrace, its location was georeferenced from images taken by the drone. A digital marker was placed on the edge of the marine terrace. This marker was used as a geo-reference in the DEM and the elevation data were extracted with the aid of an "Identify tool." The same methodology was used to extract the height data of the highest point of the adjoining spit.

### 3.3. Hydraulic Model

With determination of specific gravity based on the value of 2.3 g/cm$^3$ for andesite, a hydraulic model may be applied to predict the energy needed for the erosion of joint-bound blocks from a rocky shoreline and their subsequent transfer to an adjacent coastal boulder deposit as a function of wave impact. Andesite is a volcanic rock that forms from surface flows with variable thicknesses and a propensity to develop vertical fractures. These factors regulate the size and general shape of blocks loosened by erosion in the cliff face. Herein, two formulas are applied to estimate the magnitude of storm waves against joint-bounded boulders derived, respectively, from Equation (36) in the original work of Nott [18] (Equation (1)) and from an alternative formula that uses the velocity equations of Nandasena et al. [19] as applied by Pepe et al. (2018) [20] to estimate wave heights (Equation (2)):

$$H_S = \left( \frac{\left( \frac{\rho_s - \rho_w}{\rho_w} \right) a}{C_l} \right) \tag{1}$$

$$Hs = \frac{2\left(\frac{\rho_s - \rho_w}{\rho_w}\right).\ c\ .\ [\cos F + (\mu_s\ .\ sin\ F)]}{\frac{c_1}{100}} \tag{2}$$

where $H_s$ is the maximum height of the storm wave at breaking point; $\rho_s$ is the density of the boulder (2.3 g/cm$^3$); $\rho_w$ is the density of water at 1.02 g/cm$^3$; $a$ is the length of boulder on long axis in cm; $c$ is the length of boulder on short axis in cm; $\theta$ is the angle of the bed slope at the pre-transport location (1° for joint-bounded boulders); $\mu_s$ is the coefficient of static friction (= 0.7); and $C_1$ is the lift coefficient (= 0.178). Equation (1) is more sensitive to the length of a boulder at the long axis, whereas Equation (2) is more sensitive to the length of a boulder on the short axis. Therefore, some differences are expected in the estimates of $H_S$.

## 4. Results

### 4.1. Base Maps and Transect Lines

A set of base maps constructed on the basis of aerial photography illustrate the principal attributes of the spit (Figure 2), as located on the topographic map in Figure 1c. The massive agglomeration of loose cobbles and boulders extends for a distance of 450 m westward from the source at sea cliffs on the north side of Isla San Luis Gonzaga. A mosaic image pieced together from the aerial survey and shown in natural sunlight (Figure 2a), marks the location of seven transects with the first (T1) closest to the source of eroded andesite clasts on the north face of the island and the last (T7) most distal at the end of the spit. The surface area represented by the spit amounts to 15,600 m$^2$, of which less than 5% is obscured by plant cover dominated by the Sweet Mangrove (*Maytenus phyllanthoides*) [21]. Variations in topography (Figure 2b) reveal that the maximum elevation through the central axis of the spit rises to 3 m above mean sea level. Variations in slope direction along the divergent axes of the spit descend dominantly to the northwest (Figure 2c). Key aspects related to the layout of all transects and registered content are compiled in Table 1.

**Table 1.** Comparative data drawn from transect lines across the bar system at Bahía San Luis Gonzaga.

| Transect | Length (m) | Compass Orientation | Total Clasts Measured | Cobbles (%) | Boulders (%) | Clast Density (Clast/m) |
|---|---|---|---|---|---|---|
| 1 | 37 | 181.15° | 95 | 36 | 64 | 2.6 |
| 2 | 33 | 147.22° | 85 | 36 | 64 | 2.6 |
| 3 | 24 | 60.77° | 77 | 92 | 8 | 3.2 |
| 4 | 25 | 147.29° | 56 | 45 | 55 | 2.2 |
| 5 | 28 | 143.58° | 125 | 77 | 23 | 4.5 |
| 6 | 28 | 146.31° | 110 | 75 | 25 | 4.4 |
| 7 | 30 | 122.66° | 112 | 51 | 49 | 3.7 |
| Mean | 29 | 135.55° | 94 | 59 | 41 | 3.3 |

Average transect length amounts to 29 m and the dispositions of all but transect 3 are roughly parallel, oriented along a NW to SE trend. Transect 3 follows an orientation roughly 90° out of phase with the others, trending NE to SW. The average density of cobble and boulder clasts measured per transect is substantial at 94 with an average spacing of 3.3 clasts per meter. Transect 3 records the fewest boulders compared to all other transects at less than one in 10. Overall, the dominance of boulders over cobbles is greatest in transects 1 and 2 located most proximal to the source rocks at the beginning of the spit at a ratio 2:1. That ratio falls closer to parity between cobbles and boulders in transect 4 diagonal to the spit roughly midway along its length. Farther out along the spit in transects 5 and 6, the ratio of boulders to cobbles is 1:3. Near the tip of the spit (Figure 2a), transect 7 is the most distal from the source of eroded clasts and reflects a modest return in the relationship between boulders and cobbles at parity.

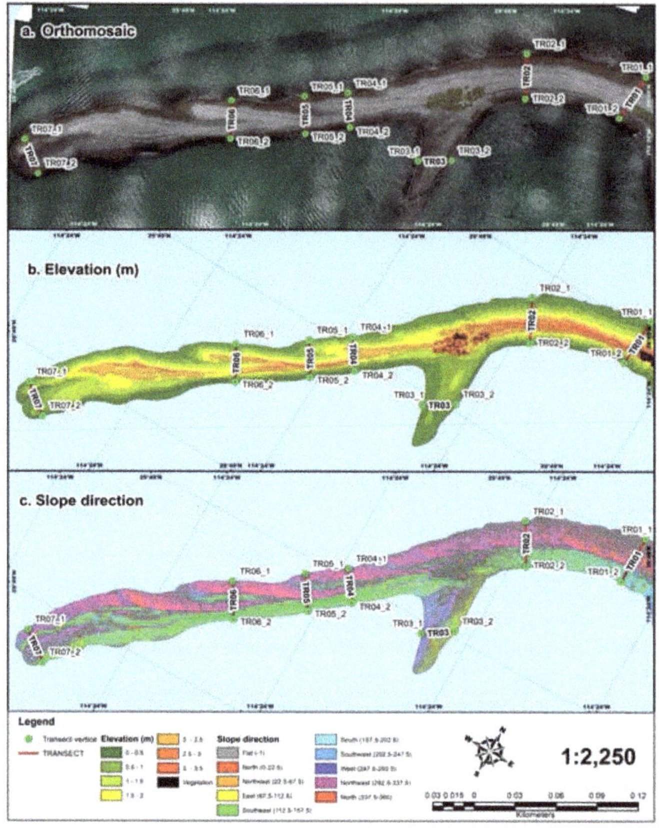

**Figure 2.** Base maps for the unconsolidated spit off the northeast end of Isla San Luis Gonzaga; (**a**) Orthophoto mosaic under natural light showing the position of transects 1 to 7; (**b**) Orthophoto color-coded map showing variations in elevation above mean sea level; (**c**) Color-coded map showing variations in slope direction.

### 4.2. Source of Joint-bound Blocks

Equations (1) and (2) from Nott (2007) and Pepe et al. (2018) [18,20] are specific to wave energy applied at the source against joint-bound blocks exposed in rocky shorelines. All materials subsequently transferred to the 450 m long spit at San Luis Gonzaga originated due to wave erosion against andesite cliffs exposed on the north side of the island. The rocky shore extends EW for almost a kilometer, rising topographically to 60 m, or more (Figure 1c). Sea cliffs in the area nearest the spit exhibit andesite flows with bedding planes that dip at a high angle to the west with irregular joints perpendicular to bedding planes (Figure 3a,b). Large blocks of andesite only crudely rounded by abrasion occupy the intertidal zone on a wave-cut platform at the side of an uplifted marine terrace. Fresh material in the supratidal zone lacks the darker tone of blocks colored by organic growth.

### 4.3. Andesite Specific Gravity

Five samples of andesite from the north shore of Isla San Luis Gonzaga yielded a range of values for specific gravity between 2.26 and 2.34 gr/cm$^3$. The samples ranged in weight between 215 and 620 gm and were displaced between 151 and 271 mL of water. The mean value calculated from the samples amounts to 2.3 gr/cm$^3$ and this value was uniformly applied to Equations (1) and (2) in estimation of wave heights provided in Tables A1–A7.

**Figure 3.** Rocky shore and intertidal zone at cliffs adjacent to the marine terrace and near the start of the 450-m long spit on Isla San Luis Gonzaga; (**a**) Oblique view with housing of meter tape for scale; (**b**) Head-on view showing tilted andesite flows in the background.

### 4.4. Comparative Variation in Clast Shapes

Raw data on boulder size in three dimensions collected from each of the seven transects are available in Appendix A (Tables A1–A7). Due to the wealth of data collected in the field, the size of Tables A1 and A2 is limited to a representative sample based on 50% of the boulders. Table A3 is the smallest, because few boulders were encountered, all of which are included. Likewise, all boulders from transects 4 to 7 are enumerated in Tables A4–A7. With regard to shape, points representing individual clasts (including smaller cobbles) are shown grouped by transect and plotted on a set of Sneed–Folk triangular diagrams (Figure 4a–g). The spread of points across all seven of the plots is remarkably consistent, showing a strong similarity in the variation of shapes from one transect to another. It is seldom that points fall into the upper-most triangle, which represents an origin from a cube-shaped endpoint. The majority of points from all seven plots falls within the middle part of the two tiers below the top triangle. Those points clustered at the core of any given triangular plot are representative of clasts for which two dimensions are closer in value than the third measured along the shortest axis. However, a significant portion of points falls into the middle-right and lower-right domains of the field, which signifies a tendency for development of elongated shapes eroded from bar-shaped blocks of source rock. The composite slope of points across all plots from the few in the topmost triangle to those in the lower right corner of the field demonstrates a tendency for development of moderately oblong shapes. Rarity of points in the middle bottom tier and complete absence of points in the lower-left corner of triangular plots indicates that the wave-eroded material from the parent sea cliffs excludes plate-shaped blocks.

### 4.5. Comparative Variation in Clast Size

Clast size is conveniently plotted on bar graphs as a function of maximum length based on the original data (see Tables A1–A7 for boulders). Transect 1 (Figure 5) is the station physically closest to the bedrock source and, in principal, is expected to reflect the highest proportion of boulders.

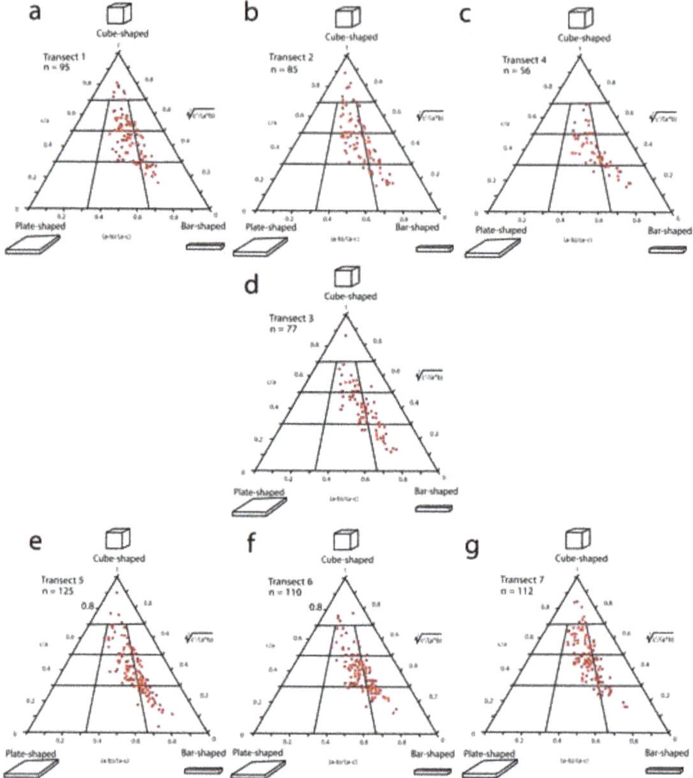

**Figure 4.** Set of triangular Sneed–Folk diagrams used to appraise variations in cobble and boulder shapes; (**a**) Trend from transect 1 closest to sea cliffs at the source of the clasts; (**b**) Trend from transect 2; (**c**) Trend from transect 4; (**d**) Isolated trend from transect 3 with a markedly different orientation from all other transects; (**e**) Trend from transect 5; (**f**) Trend from transect 6; (**g**) Trend from transect 7 most distant from sea cliffs at the common source of all clasts in the system.

Conversely, transect 7 is most distal from the bedrock source and is expected to show a higher proportion of cobbles. Groupings separated in bins at intervals of 5 cm are arrayed in histograms stacked to show differences in size range between 6 cm and 90 cm among transects 1, 2, and 4 (Figure 6a–c). Data for transect 1 (Figure 6a) registers the highest concentration of small boulders. Transect 3 projects as a side spur on the southeast side of the spit (Figure 7). It is excluded from this analysis due to the relative scarcity of boulders. Each of the three graphs in Figure 6 delineates the boundary between cobbles and boulders with a dashed line. They are consistently skewed with the highest percentage of clasts at or around the border between the largest cobbles and smallest boulders. The ratio between boulders and cobbles remains steady at 2:1 in transects 1 and 2 but is closer to parity in transect 4. Compared to the proximal transects, the more distal transects 5 to 7 (Figure 8a–c) exhibit a marked shift in skewness to a numerical domination by cobbles. In transects 5 and 6, the relationship between boulders to cobbles is roughly consistent dropping to a ratio of 1:3. However, data from transect 8 at the distal end of the spit (Figure 8c) records a ratio at parity. In all cases among the six transects represented by bar graphs, the extreme outlier of large boulders occurs at measured lengths between 76 cm and 85 cm.

**Figure 5.** Transect 1 (dashed line) features the highest concentration of small and intermediate size boulders. The survey line passes through the spit about 40 m beyond the marine terrace at the NE corner of Isla San Luis Gonzaga. Superposition of white arrows defines the outer lip of the marine terrace elevated 8.5 m above sea level.

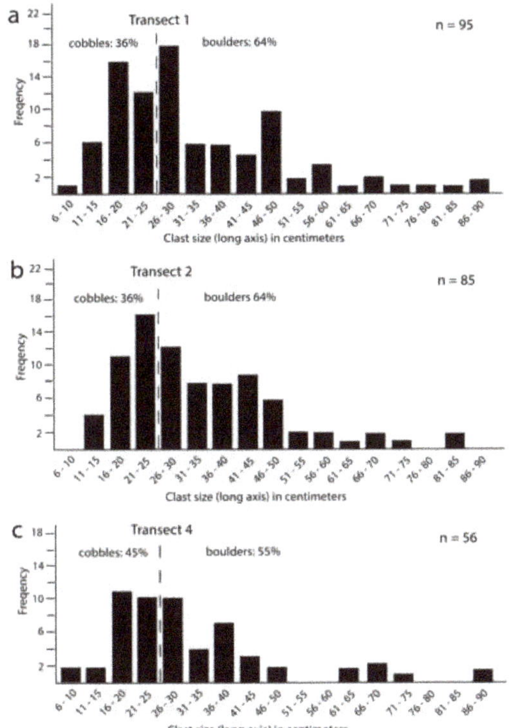



**Figure 6.** Set of bar graphs used to appraise variations in maximum boulder length from the three transects with similar orientations closest to the source of coastal erosion; (**a**) Bar graphs from transect 1; (**b**) Bar graphs from transect 2; (**c**) Bar graphs from transect 4. Transect 3 is excluded from this treatment on account of its deviant orientation.

**Figure 7.** Transect 3 (dashed line) follows a side spur off the main spit with a compass orientation 90° out of phase with other transects. Boulders are uncommon, but the largest in the foreground exceeds 50 cm in diameter. The spur is notably isolated from the opposite side of the spit by a dense thicket of Sweet Mangrove (*Maytenus phyllanthoides*) and is largely submerged during high tide.

*4.6. Variation in Transect Profile Elevations*

Data recovered from the seven transects in this study (Figure 2) also include the requisite information for construction of individual elevation profiles. The highest elevation determined at the crest of any single transect amounts to no more than 3 m above mean sea level. Transects 1 to 3 were found to conform to profiles that rise evenly to peak elevation at or near the center of the transect from opposite ends and are not illustrated. In contrast, transects 4 to 6 (Figure 9) exhibit profiles with a marked depression at variable positions along the line. Transect 4 (Figure 9a) registers a modest decline in elevation of 50 cm across the central 5 m of the line. Transect 5 (Figure 9b) shows a similar dip but is skewed much closer to the NW end of the transect. Transect 6 (Figure 9c) exhibits the deepest depression midway through the line with a drop of about a meter. Transect 7 also features a bifurcated elevation profile, but it is not shown at the same scale because it represents a much lower overall value in maximum elevation at only 1.2 m. Among the seven transects, transect 3 (see Figure 7) registered the lowest elevation rising to only a half meter above sea level.

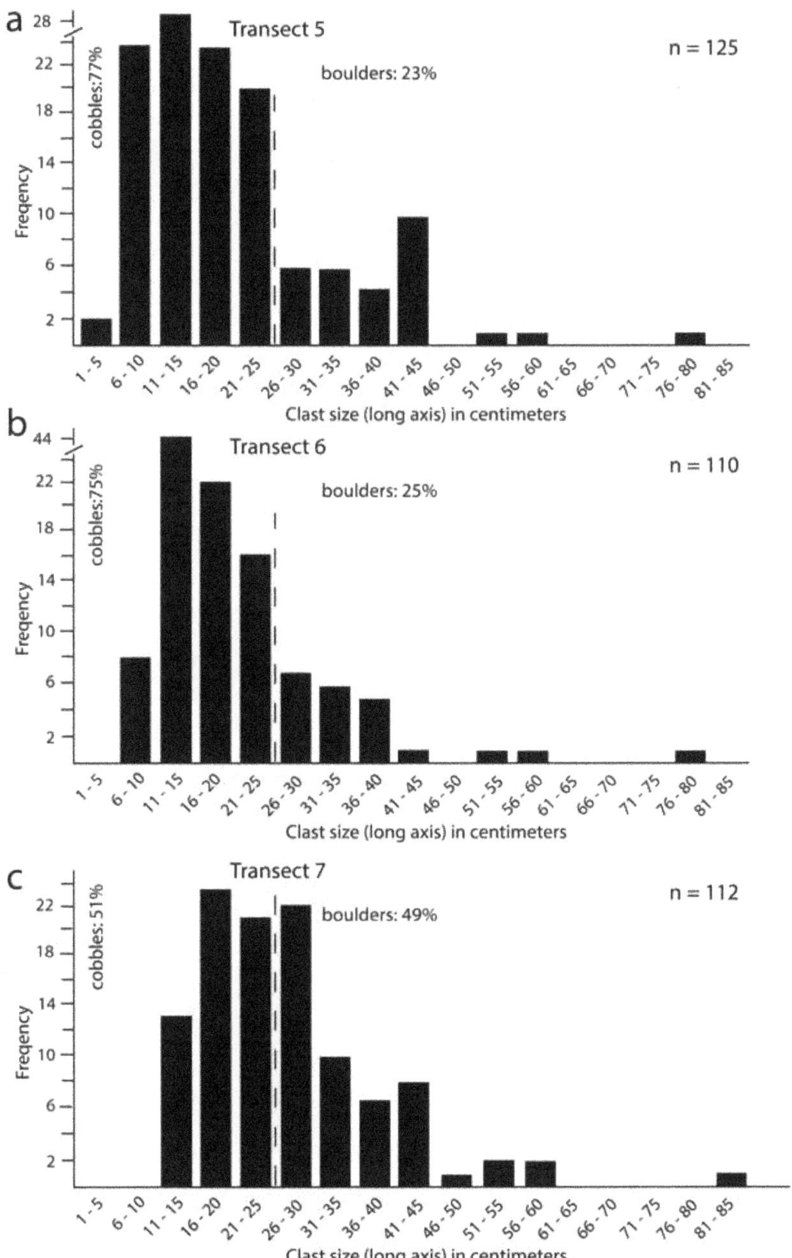


**Figure 8.** Set of bar graphs used to appraise variations in maximum boulder length from the three transects with similar orientations most distal from the source of coastal erosion; (**a**) Bar graphs from transect 5; (**b**) Bar graphs from transect 6; (**c**) Bar graphs from transect 7.

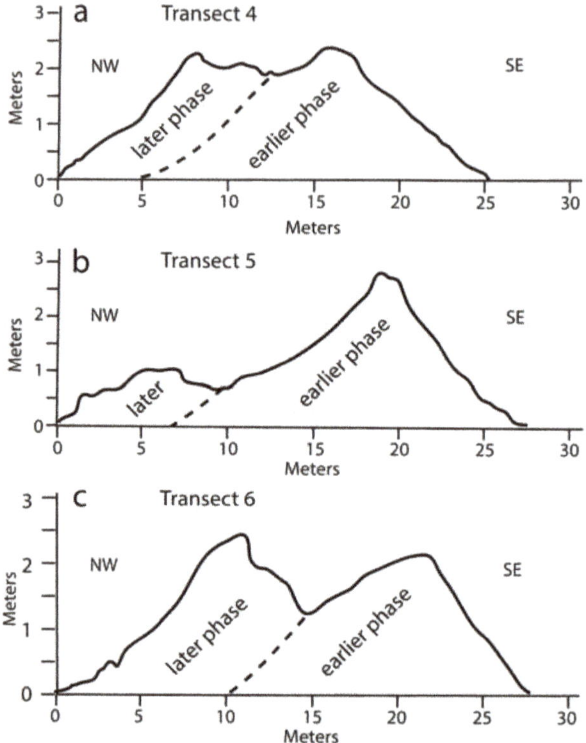



**Figure 9.** Comparison of selected transverse profiles in the distal part of the spit showing separation of different phases in the overall deposit. Transects 1–3 and 7 lack such distinct features; (**a**) Elevation profile through transect 4; (**b**) Elevation profile through transect 5; (**c**) Elevation profile through transect 6. See Section 3 for source of the elevation data.

The long view over the spit's axis to the NE from a location on transect 4 (Figure 10) captures the nature of the depression cutting diagonally through the more proximal part of the structure. Thus, the transverse depression in transect 4 is defined by a pair of longitudinal bars that distinctly separate one side of the spit from the other with an intermediate trough. From this vantage, it may be realized that the spit underwent a growth history in sequential phases. Following from this insight, the ramification is that the Sneed–Folk plots in Figure 4c,e,f, as well as the bar graphs in Figures 6c and 7 represent consecutive phases of development that occurred over time. In contrast, the elevation profiles for transects 1–3 that conform to a simple arc in relief, imply that the Sneed–Folk plots and bar graphs relevant to those transect samples reflect more unitary slices temporal development.

### 4.7. Biological Data from Encrusted Boulders

From place to place, disarticulated bivalve shells (including *Anadara grandis* and *Megapitaria squalida*) appear sporadically among the mixed cobbles and boulders forming the spit. The degree of breakage in some of the larger shells indicates a higher level of energy was entailed in their delivery to the spit, more than from normal conditions related to tidal flux. More significant are examples of large boulders encrusted by the common oyster (*Ostrea palmula*) that thrives in intertidal waters throughout the Gulf of California [22]. A block with a long axis of 70 cm found near transect 7 hosted more than two dozen oysters on its upper surface (Figure 11a) before it was transported onto the spit and left upside down. In contrast to the inarticulate shells found elsewhere separately as disarticulated shells, the oysters are encrusted in growth position and many remain articulated (Figure 11b). The biological inference is that the boulder originally sat submerged in the shallow water off to the side of the spit and was subsequently carried onto the spit by a storm of sufficient strength to lift an andesite block weighting as much as 80 to 100 kg (see range of weights in Table A7). The pristine condition of the oyster shells suggests that the storm was a relatively recent event. Weaker storms are capable of transporting cobles and smaller boulders during earlier events and may have shifted the offshore position of the block until the next major storm moved it onto the spit.

**Figure 10.** View across transect 4 northeast toward the connection with Isla San Luis Gonzaga, showing a distinct swale between two different phases of the cobble-boulder deposit. Note the raised Pleistocene marine terrace in the far distance (white arrows). The inner wall of the marine terrace is oriented NS.

**Figure 11.** Contemporary oysters (*Ostrea palmula*) encrusted on a boulder near transect 7; (**a**) Boulder as first encountered on the spit (meter stick for scale); (**b**) Same boulder overturned for better view of encrusting oysters, approximately 7 cm in shell length. Note: many of the oysters remain articulated.

### 4.8. Storm Intensity as Function of Estimated Wave Height

Average boulder sizes and maximum boulder sizes from all seven transects are summarized in Table 2, whether or not the respective Sneed–Folk diagrams and bar graphs (Figure 4, Figure 6, and Figure 8) reflect solitary events or an agglomeration of phased events. These data are a prerequisite for comparison of results estimating wave heights first directed against joint-bound blocks on the andesite rocky shore as derived from the Nott formula [18] and the subsequent formula applied by Pepe et al. [20]. There exists a general trend in reduction of average boulder size from transect 1 to transect 5 more than halfway along the Gonzaga spit from the proximal source. Thereafter in transects 6 and 7, there occurs an increase in average boulder size. The estimated mean wave height necessary to transport those boulders extracted from the rocky shore amounts to 2.7 m according to the Nott formula. However, at 73 cm the mean maximum boulder length derived from all seven transects yields a value almost twice the diameter for the average of all averages (Table 2). Moreover, the estimated mean weight of the largest single boulder from each transect amounts to 200 kg, which is more than 4.5 times the average weight computed from all boulders surveyed in the seven transects. The estimated wave height needed to shift the largest boulder in transect 7 amounts to nearly 6 m according to the Nott equation, but half that compared with the Pepe equation. However, results based on the Pepe equation for the largest boulders yield a higher wave height in four out of seven transects. In general agreement with the biological inference from the oyster-encrusted block in Figure 11, the critical insight from these comparative data is that the average impact from smaller waves is sufficient to move smaller boulders, but only the largest waves are sufficient to move the largest boulders.

**Table 2.** Summary data from Appendix A (Tables A1–A7) showing maximum bolder size and estimated weight compared to the average values for sampled boulders from each of the transects together with calculated values for wave heights estimated as necessary for CBD mobility. EWH = estimated wave height.

| Transect | Number of Samples | Mean Boulder Size (cm$^3$) | Mean Boulder Weight (kg) | Estimated Mean Wave ht. Nott [18] (m) | Max. Boulder Size (cm$^3$) | Max. Boulder Weight (kg) | Max. EWH Nott [18] (m) | Max. EWH Pepe et al. [20] (m) |
|---|---|---|---|---|---|---|---|---|
| 1 | 32 | 43.5 | 81 | 3.0 | 90 | 628 | 6.3 | 7.6 |
| 2 | 33 | 40 | 59 | 2.8 | 64 | 139 | 3.7 | 5.6 |
| 3 | 6 | 36 | 12.75 | 2.7 | 61 | 35 | 4.3 | 1.3 |
| 4 | 32 | 40 | 36 | 2.8 | 75 | 191 | 5.3 | 5.6 |
| 5 | 29 | 25 | 29.5 | 2.6 | 57 | 166 | 4.0 | 6.1 |
| 6 | 20 | 35 | 25 | 2.4 | 80 | 190 | 5.6 | 3.7 |
| 7 | 30 | 38 | 47 | 2.6 | 84 | 122 | 5.9 | 3.2 |
| Mean | 26 | 37 | 43 | 2.7 | 73 | 200 | 5.0 | 4.7 |

## 5. Discussion

### 5.1. Phased Development During Holocene Time

As projected by the orthophoto mosaic in Figure 2a and supplemented by transect elevation profiles derived from seven transects, the overall layout of the unconsolidated cobble-boulder spit at Isla San Luis Gonzaga suggests an interpretation of growth through multiple phases during Holocene time. A starting point in post-Pleistocene time is supported by the physical connection of the spit to the island adjacent to an uplifted marine terrace (see photo in Figure 5). The terrace truncates the northwest corner of the island and its outer lip rises 8.5 m above the base level of the spit. Marine erosion on the terrace ceased prior to initiation of the spit around the present sea level. The recessed terrace flat is relatively clean, showing that the Pleistocene sea cliff at the rear of the terrace was not the parent source of eroded clasts contributing to the spit. Instead, the yet active source appears along the modern sea cliffs that stretch across the northern part of the island. At the proximal end of the spit, sea cliffs on the north exposure rise steeply up to 60 m in height (Figure 1c). Headward erosion of a modern wave-cut platform cuts into the base of adjoining sea cliffs that provide the copious raw materials derived from joint-bound blocks (Figure 3). Large blocks are first smoothed by wear in the surf and subsequently transferred by storm currents in a SW direction to the spit. Undermining of the sea cliff by storm action also contributes to rock falls from higher in the exposure.

Creation of the 450-m long spit is interpreted as having evolved over the last 10,000 years during a succession of episodic storm events outlined in Figure 11. Progradation of the spit at the outset between transects 1 and 2 follows a linear pattern that left a consistent profile tracing a single, central rise in elevation roughly midway between side margins (Figure 2b). The next phase of construction entailed a curvature to the south that terminated beyond the position of transect 3 (Figure 12a). The profile across transect 3 reflects a low median rise in elevation (see Figure 7). A dense thicket of vegetation north of transect 3 (Figure 2a) signals this phase of development terminated as a side spur that ceased to receive fresh material and became isolated from the rest of the structure. Resumption of deposition with a linear extension to the SW pushed a narrow lobe of the spit beyond transect 4 (Figure 12c), presumably due to a change in storm dynamics. The pair of topographic bars that form parallel swales in transects 4 and 5 (Figure 9a,b) mark the further expansion of the spit to the SW with the more easterly bar deposited during an earlier storm event and the adjoining bar amalgamated alongside during a later event. Physical compression of the two storm events resulted in expansion of the spit's width across transects 4 and 5.

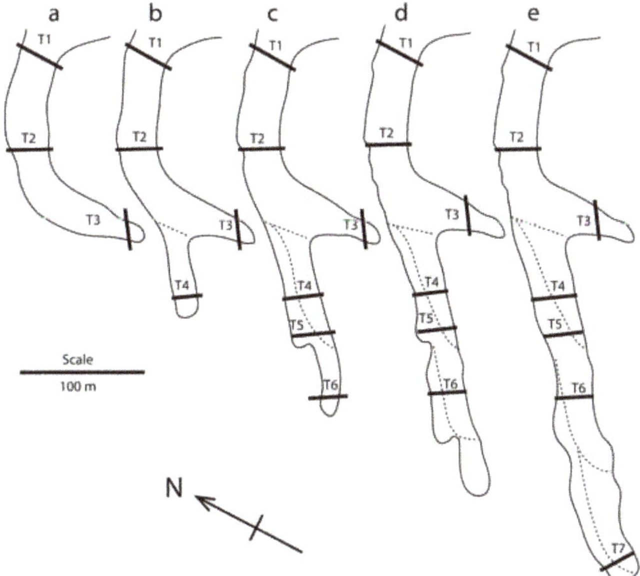


**Figure 12.** Interpretation of temporal development through Holocene time with study transects marked as reference points: (**a**) phase one; (**b**) phase two; (**c**) phase three; (**d**) phase four; (**e**) phase five. Dashed lines mark boundaries between successive additions to the structure through time.

A similar sequence of staggered events took place with further extension of the spit to the SW beyond transect 6 (Figure 12d). During the earlier phase of expansion in this area, the spit was narrower but doubled in width with amalgamation of the adjoining later phase. The pattern repeated itself, once again, under progradation of the spit beyond transect 7 (Figure 12e) to reach the present terminal length of 450 m. The number of discrete storm events that contributed to the phased growth of the spit is difficult to estimate, but the presence of boulders across all transects makes clear that major storms were involved.

The incremental rate of the structure's progradation from the bedrock source on Isla San Luis Gonzaga is difficult to calculate with any accuracy, but an estimate can be offered on the assumption that phased growth resulted from storms of hurricane strength that occurred as often as every 100 years. This metric is suggested in reference to the popular notion of recurrent events outside the living memory of a long human lifetime. Holocene mega-storms with a measurable recurrence between 100 and 300 years are tested on the basis of coastal storm ridges that incorporate coral heads dated by isotope chronology [5]. Recurrence of mega-storms on a centennial scale also is suggested by the isotope chronology of speleothem deposits from caves commonly impacted by landfall of tropical cyclones [23]. No such method of absolute dating is possible with regard to the storm-deposited andesite boulders from Isla San Luis Gonzaga. The rocks are dated by radiometric means, but the dates so derived represent the age when lava flows solidified during the Miocene [13] and not the time of coastal erosion.

Storms of hurricane intensity reaching Isla San Luis Gonzaga would have entailed wave heights at or exceeding 5 m based on equations from Nott [18] and Pepe et al. [20]. Such waves would impact the island's north shore following a pattern of cyclonic rotation with northward travel. Extraction of large andesite blocks from the base of the sea cliffs (Figure 3) was influenced by hydraulic wedging of joints and partings in the layered andesite during wave impact. The subsequent transport of boulders to the proximal end of the spit was driven by vigorous storm-generated currents. Successive big storms may have moved the larger boulders piece-meal for shorter distances during each event. Over the 10,000-year span of the Holocene, 100 super-storms may have reached the upper Gulf of California. Such a reckoning is reasonable, given the spit's geometry and various internal boundaries demarcated by amalgamated bars (Figure 12). An episodic growth rate between 7 and 8 m/century is informed by the total length of the various components adding up to a composite total of 750 m. Whereas each cobble and boulder in the construction is ultimately traced to the sea cliffs on the island's north shore, the residency time of large blocks resting offshore also must be considered. Thus, large andesite blocks of considerable weight may sit offshore near the distal end of the spit for some time during which biological encrustations accumulate (Figure 11), before the block is shifted from inter-tidal waters up onto the spit during a major storm.

*5.2. Inference from Historical Hurricanes*

Hurricane Odile entered the Gulf of California as a Category 4 storm in September 2014 and its economic impact is regarded as one of the most destructive events to affect the peninsular state of Baja California Sur, having caused more than 1654 million USD in damage to coastal infrastructure largely as a result of high winds [1]. According to a subsequent assessment by Gross and Magar (2020) [3], the same storm had the capacity to damage tidal-energy transformers due to wind-driven waves if such mechanisms were in place as part of the energy grid serving the state of Baja California at the far end of the gulf in the north. No such infrastructure presently exists in the region, but the study offers a cautionary warning about the potential for such damage even in the upper part of the gulf where hurricanes typically degrade to tropical storms. The last major storm to have crossed the upper gulf was Hurricane Kathleen in September 1976 as a Category 3 hurricane [21]. The passage of time since that event is considerable in terms of human memory and may be thought of as a once in a lifetime event by local residents living around Bahía San Luis Gonzaga.

The peninsular region of western Mexico was spared a potentially catastrophic event during the 2015 storm season, when Hurricane Patricia formed as a Category 5 storm with wind speeds up to 346 km/h off the Mexican mainland well south of the Baja California peninsula. That storm still holds the record as the most powerful hurricane to have originated in the eastern North Pacific Ocean [24]. Only the storm's outer most bands brushed across the opening to the Gulf of California, but the center of the storm made landfall on the Mexican mainland after an unexpected turn to the east. Had Hurricane Patricia tracked into the Gulf of California, it was certain to have caused more damage to coastal infrastructure than Hurricane Odile during the previous season. The occurrence of these two powerful storms, one after the other in subsequent years, raises the question of accountability for storms attributed to 100-year events. With growing conditions of global warming now experienced around the world, the increase of such events is sure to be compressed in frequency as suggested by re-evaluations of storm data on a decadal basis [25]. Under such circumstances, the study of coastal boulder deposits and their development in recent geologic time offers a window on natural processes of interest not only to geomorphologists but also to engineers challenged to design infrastructures better suited to withstand more intense storms and any future rise in sea level.

### 5.3. Comparison with Other Gulf of California Deposits

Study of coastal boulder deposits (CBDs) and related boulder barrier deposits (BBDs) in the Gulf of California is a pursuit that has gained momentum during the last few years but, heretofore, only in the lower Gulf of California [9–11]. This contribution is the first to focus on the upper Gulf of California, a region generally thought to be impacted much less from the severity of hurricanes than the more southern parts closer to the Tropic of Cancer. Earlier work looked at limestone boulders pulled by storm action from Pliocene sea cliffs on Isla del Carmen, where a line of eroded boulders sits high on a marine terrace [9]. In this regard, the Carmen example conforms to a classic CBD pealed back from the front of a sea cliff by storm-related overwash. The Isla San Luis Gonzaga example in this study corresponds to the formation of boulder bars, where large blocks of andesite are leveraged from joint-bound layers at the base of sea cliffs and transported laterally by storm-generated currents where they accumulate as spits. Compared to previous examples in the southern Gulf region also formed by volcanic boulders derived from joint-bound blocks [11,12], the Isla San Luis Gonzaga structure is larger and more complex with components that can be differentiated as having accumulated during discrete storm events. For example, the earliest phases of extension ending beyond transect 3 off Isla San Luis Gonzaga (Figure 12a) are comparable to the half-ring structure constructed from rhyolite boulders at Ensenada Almeja [10]. The Almeja structure traces a pattern of clockwise extension stretching in an arc over a distance of about 250 m from the original bedrock source with joint-bound blocks. That pattern is replicated in the same dimensions by the earliest phases of construction at Isla San Luis Gonzaga. In both examples, fewer boulders are found on top of the bar closer to its termination, suggesting that wave energy declined with distance from the proximal end of those structures at the bedrock source.

After isolation of the spur leading to transect 3 from the rest of the spit at Isla San Luis Gonzaga (Figure 12b), new growth was extended in a strictly linear fashion similar to the unidirectional extension of bars across the front of Puerto Escondido in the lower Gulf of California [11]. There, two bars follow a linear fault trace with a small island connected in between. Not including the islet, the two bars extend for a length of 250 and 140 m, respectively. By comparison, the apparent length of the Gonzaga spit amounts to 450 m. However, development of parallel bars that became progressively attached to the core from one side, the total length of component parts including duplications amounts to a distance of 750 m. As such, the Gonzaga spit is one of most complex structures of its kind formed by loose cobbles and boulders in the entire Gulf of California region. Based on shape analyses from the various localities throughout the gulf region, there is little difference in the blocks of rock that were worn by marine erosion from limestone, rhyolite, and andesite into oblong cobbles and boulders. The factor that makes results of the Isla San Luis Gonzaga study so important is the apparent correspondence between extraordinary size, complexity, and the length of time necessary for development of BBDs. Given the weight of the largest boulders distributed across much of the Isla San Luis Gonzaga spit, it is apparent that their transport occurred during episodic hurricanes of high intensity. With greater attention to examples of CBDs throughout the Gulf of California and the expectation that the frequency of future hurricanes is likely to increase, it is proposed that known sites [9–11] including the Isla San Luis Gonzaga spit be monitored for changes once a selection of the largest boulders is tagged for reference.

### 5.4. Comparison with island deposits in the North Atlantic

Studies on CBDs on islands in the North Atlantic Ocean show much promise for similar research applying the same kind of analyses performed in the Gulf of California. An identical program using the same techniques of analyses for boulder shapes and sizes was conducted on modern and Pleistocene

CBDs on Santa Maria Island in the Azores [26]. In particular, the same equations have been applied to estimate wave heights during the Late Pleistocene (Marine Isotope Substage 5e) on basalt-dominated shores dated by fossils to approximately 125,000 years ago. Future studies that incorporate Pleistocene fossils may be expected in other Atlantic archipelagos such as the Canary and Cape Verde Islands.

## 6. Conclusions

Satellite tracking for hurricanes and typhoons has improved the ability of meteorologists to gather and analyze data on changes in atmospheric pressure, wind speed, and other factors that make statistical predictions based on recurrent circulation patterns increasingly accurate on a global basis [24]. Historically, few of the big storms generated in the NE Pacific Ocean actually deviate in direction to enter Mexico's Gulf of California from their point of origin farther south off the western mainland [1]. Those that do are generally found to rapidly deteriorate from hurricane strength to that of lesser tropical storms [21]. In contrast, the recent geologic record from the Holocene expands the time frame for evaluation of storms equilibrated with deposits of eroded shoreline materials and their bearing on coastal geomorphology. The present study on the complex boulder bar associated with Isla San Luis Gonzaga in the upper Gulf of California permits the following conclusions:

(1) Coastal boulder deposits and related boulder bars are described from the lower Gulf of California that experience episodic hurricanes, but the Isla San Luis Gonzaga spit composed of unconsolidated materials in the upper Gulf of California is larger and more complex than any features previously studied in the Lower Gulf of California.

(2) All cobbles and boulders entrained in the 450-m long structure are derived from andesite sea cliffs with joint-bound blocks on the north side of Isla San Luis Gonzaga, bypassing a well-defined marine terrace from the Late Pleistocene.

(3) Extensive data on variations in shape collected from seven transects across the structure show that clasts are mostly elongated in configuration. Data on variations in clast size from the same transects indicate changing ratios between boulders and cobbles that fall from 3:1 in favor of boulders to more equitable proportions in a progression toward the end of the structure.

(4) Despite a general decrease in the boulder population along the length of the structure, large boulders continue to be present even in the most distal parts of the structure and the estimated wave heights required to move those blocks into place entails an average height of 5 m.

(5) The overall complexity of the structure includes parallel bars that form side-by side as distinct swales through the distal half of the spit. These are interpreted as discrete additions from episodic storms. Assuming the structure began to form at the start of the Holocene, there was sufficient time for multiple mega-storms to reach the upper Gulf of California at a general rate of one per century to yield a possible growth rate between 7 and 8 m/century.

(6) Civil engineers evaluate hurricane damage to infrastructure as due to storm wind speed ashore and the impact of wave surge in coastal waters, but the geomorphology of coastal boulder deposits and related boulder bars provides another means to assess the potential for storm risk in regions like Mexico's Baja California peninsula and the adjacent Gulf of California.

(7) The concept of the 100-year storm as an event exceeding human memory at any point in time has changed during the historic rise in global warning experienced over the last two decades. It is not a question of if, but rather when, the next hurricane comparable to the Category 5 Patricia in 2014 will strike the gulf coast of Mexico's Baja California peninsula. As a follow-up to the region's various studies on coastal boulder deposits, consideration must be given to a monitoring program whereby some of the largest boulders are tagged to test movements after the next big storm.

**Author Contributions:** Conceptualization, R.G.-F. and J.L.-V.; methodology, R.G.-F. and M.A.S.R.-d.R.; software and validation, Á.R.H.-G.; formal analysis, writing, and draft preparation, M.E.J. All authors have read and agreed to the published version of the manuscript.

J. Mar. Sci. Eng. **2020**, 8, 525

**Funding:** We are grateful to the Universidad Autónoma de Baja California for its support in collecting and processing the baseline information reported, herein, as part of the research project "Campaña Exploratoria para Inventariar el Patrimonio Geológico del Corredor Turístico Pluertecitos—San Luis Gonzaga Ensenada B.C., México" Programmatic number 388, supported in the *20va Convocatoria de Apoyo a Proyectos de Investigación*.

**Acknowledgments:** The following faculty and students from the Univesidad Autónoma de Baja California are thanked for their participation during field work on Isla San Luis Gonzaga: Miguel Agustín Téllez-Duarte, Cesar Omar Luna-Miyaki, and Josseline Michelle Crus-Pérez. Five readers contributed peer reviews with useful comments that led to the improvement of this contribution. Application of the equations used in this study is the sole responsibility of the authors.

**Conflicts of Interest:** The authors declare no conflicts of interest.

## Appendix A

**Table A1.** Quantification of boulder size, volume, and estimated weight from selected coastal bar samples through transect 1 at Bahía San Luis Gonzaga. EWH = estimated wave height. The density of Baja Californian andesite at 2.3 gm/cm$^2$ is applied uniformly in order to calculate wave height for each boulder.

| Sample | Long Axis (cm) | Intermediate Axis (cm) | Short Axis (cm) | Volume (cm$^3$) | Adjust to 75% | Weight (kg) | EWH Nott [18] (m) | EWH Pepe [20] (m) |
|---|---|---|---|---|---|---|---|---|
| 1 | 80 | 55 | 50 | 220,000 | 165,000 | 421 | 5.6 | 8.0 |
| 4 | 42 | 28 | 13 | 15,288 | 11,466 | 29 | 3.0 | 2.1 |
| 6 | 68 | 60 | 53 | 216,240 | 162,180 | 414 | 4.8 | 8.4 |
| 8 | 32 | 22 | 10 | 7040 | 5280 | 13 | 2.3 | 1.6 |
| 10 | 38 | 27 | 16 | 16,416 | 12,312 | 31 | 2.7 | 2.5 |
| 12 | 50 | 29 | 10 | 14,500 | 10,875 | 28 | 3.5 | 1.6 |
| 14 | 46 | 30 | 12 | 16,560 | 12,420 | 32 | 3.2 | 1.9 |
| 16 | 40 | 28 | 11 | 12,320 | 9240 | 24 | 2.8 | 1.8 |
| 19 | 50 | 28 | 25 | 35,000 | 26,250 | 67 | 3.5 | 4.0 |
| 21 | 60 | 45 | 21 | 56,700 | 42,525 | 108 | 4.2 | 3.3 |
| 23 | 90 | 76 | 48 | 328,320 | 246,240 | 628 | 6.3 | 7.6 |
| 27 | 56 | 40 | 12 | 26,880 | 20,160 | 51 | 3.9 | 1.9 |
| 29 | 44 | 30 | 15 | 19,800 | 14,850 | 38 | 3.1 | 2.4 |
| 31 | 57 | 29 | 17 | 28,101 | 21,076 | 54 | 4.0 | 2.7 |
| 35 | 25 | 13 | 7 | 2275 | 1706 | 4.4 | 1.8 | 1.1 |
| 37 | 27 | 22 | 11 | 6534 | 4900 | 12.5 | 1.9 | 1.8 |
| 40 | 53 | 48 | 15 | 38,160 | 28,620 | 73 | 3.7 | 2.4 |
| 42 | 28 | 23 | 15 | 9660 | 7245 | 18.5 | 2.0 | 2.4 |
| 44 | 40 | 24 | 20 | 19,200 | 14,400 | 37 | 2.8 | 3.2 |
| 46 | 27 | 21 | 20 | 11,340 | 8505 | 22 | 1.9 | 3.2 |
| 49 | 32 | 30 | 24 | 23,040 | 17,280 | 44 | 2.3 | 3.8 |
| 52 | 35 | 28 | 20 | 19,600 | 14,700 | 37 | 2.5 | 3.2 |
| 56 | 50 | 20 | 18 | 18,000 | 13,500 | 34 | 3.5 | 2.9 |
| 59 | 59 | 30 | 23 | 40,710 | 30,533 | 78 | 4.2 | 3.7 |
| 63 | 28 | 21 | 20 | 11,760 | 8,820 | 22.5 | 2.0 | 3.2 |
| 65 | 46 | 40 | 37 | 68,080 | 51,060 | 130 | 3.2 | 5.9 |
| 70 | 30 | 23 | 16 | 11,040 | 8280 | 21 | 2.1 | 2.5 |
| 73 | 44 | 35 | 16 | 24,640 | 18,480 | 47 | 3.1 | 2.5 |
| 77 | 27 | 22 | 10 | 5940 | 4455 | 11.4 | 1.9 | 1.6 |
| 80 | 41 | 35 | 22 | 31,570 | 23,678 | 60 | 2.9 | 3.5 |
| 82 | 25 | 16 | 10 | 4000 | 3000 | 7.5 | 1.8 | 1.6 |
| 87 | 25 | 14 | 4 | 1400 | 1050 | 2.8 | 1.8 | 0.6 |
| Mean | 43.6 | 31 | 19 | 42,504 | 31,878 | 81 | 3.1 | 3.1 |

**Table A2.** Quantification of boulder size, volume, and estimated weight from selected coastal bar samples through transect 2 at Bahía San Luis Gonzaga. EWH = estimated wave height. The density of Baja Californian andesite at 2.3 gm/cm$^2$ is applied uniformly in order to calculate wave height for each boulder.

| Sample | Long Axis (cm) | Intermediate Axis (cm) | Short Axis (cm) | Volume (cm$^3$) | Adjust to 75% | Weight (kg) | EWH Nott [18] (m) | EWH Pepe [20] (m) |
|---|---|---|---|---|---|---|---|---|
| 1 | 33 | 25 | 24 | 19,800 | 14,850 | 38 | 2.3 | 3.8 |
| 5 | 47 | 35 | 33 | 54,285 | 40,714 | 104 | 3.3 | 5.3 |
| 7 | 42 | 22 | 20 | 18,480 | 13,860 | 35 | 3.0 | 3.2 |
| 9 | 52 | 40 | 35 | 72,800 | 54,600 | 139 | 3.7 | 5.6 |
| 11 | 58 | 47 | 26 | 70,876 | 53,157 | 136 | 4.1 | 4.1 |
| 13 | 31 | 18 | 6 | 3348 | 2511 | 6.4 | 2.2 | 1.0 |
| 15 | 50 | 29 | 16 | 23,200 | 17,400 | 44 | 3.5 | 2.5 |
| 17 | 36 | 13 | 11 | 5148 | 3861 | 9.8 | 2.5 | 1.8 |
| 20 | 54 | 37 | 25 | 49,550 | 37,463 | 96 | 3.8 | 4.0 |
| 22 | 46 | 25 | 18 | 20,700 | 15,525 | 40 | 3.2 | 2.9 |
| 23 | 42 | 33 | 18 | 328,320 | 246,240 | 628 | 3.0 | 2.9 |
| 26 | 28 | 21 | 10 | 5880 | 4410 | 11 | 2.0 | 1.6 |
| 28 | 26 | 24 | 10 | 6240 | 4680 | 12 | 1.8 | 1.6 |
| 30 | 33 | 30 | 29 | 28,710 | 21,533 | 55 | 2.3 | 4.6 |
| 32 | 28 | 18 | 12 | 6048 | 4536 | 12 | 2.0 | 1.9 |
| 34 | 30 | 15 | 7 | 3150 | 2363 | 6 | 2.1 | 1.1 |
| 39 | 29 | 23 | 11 | 7337 | 5503 | 14 | 2.0 | 1.8 |
| 41 | 45 | 25 | 24 | 27,000 | 20,250 | 52 | 3.2 | 3.8 |
| 45 | 39 | 36 | 18 | 25,272 | 18,954 | 48 | 2.7 | 2.9 |
| 47 | 50 | 38 | 16 | 30,400 | 22,800 | 58 | 3.5 | 2.5 |
| 51 | 44 | 20 | 12 | 10,560 | 7,920 | 20 | 3.1 | 1.9 |
| 53 | 64 | 51 | 11 | 35,904 | 26,928 | 69 | 4.5 | 1.8 |
| 55 | 30 | 30 | 18 | 16,200 | 12,150 | 31 | 2.1 | 2.9 |
| 57 | 50 | 28 | 8 | 11,200 | 8400 | 21 | 3.5 | 1.3 |
| 59 | 44 | 42 | 10 | 18,480 | 13,860 | 35 | 3.1 | 1.6 |
| 63 | 43 | 23 | 13 | 12,857 | 9643 | 25 | 3.0 | 2.1 |
| 66 | 29 | 28 | 20 | 16,240 | 12,180 | 31 | 2.0 | 3.2 |
| 68 | 41 | 36 | 13 | 19,188 | 14,391 | 37 | 2.9 | 2.1 |
| 73 | 37 | 29 | 20 | 21,460 | 16,095 | 41 | 2.6 | 3.2 |
| 75 | 26 | 19 | 15 | 7410 | 5558 | 14 | 1.8 | 2.4 |
| 77 | 40 | 25 | 25 | 25,000 | 18,750 | 48 | 2.8 | 4.0 |
| 80 | 23 | 23 | 8 | 4600 | 3450 | 8.8 | 1.6 | 1.3 |
| 84 | 36 | 21 | 18 | 13,608 | 10,206 | 26 | 2.5 | 2.9 |
| Mean | 40 | 28 | 17 | 30,886 | 23,174 | 59 | 2.8 | 2.7 |

**Table A3.** Quantification of boulder size, volume, and estimated weight from selected coastal bar samples through transect 3 at Bahía San Luis Gonzaga. EWH = estimated wave height. The density of Baja Californian andesite at 2.3 gm/cm$^2$ is applied uniformly in order to calculate wave height for each boulder.

| Sample | Long Axis (cm) | Intermediate Axis (cm) | Short Axis (cm) | Volume (cm$^3$) | Adjust to 75% | Weight (kg) | EWH Nott [18] (m) | EWH Pepe [20] (m) |
|---|---|---|---|---|---|---|---|---|
| 1 | 31 | 16 | 6 | 2976 | 2232 | 5.7 | 2.2 | 1.0 |
| 17 | 61 | 38 | 8 | 18,544 | 13,908 | 35 | 4.3 | 1.3 |
| 19 | 31 | 14 | 9 | 3906 | 2930 | 7.5 | 2.2 | 1.4 |
| 40 | 40 | 27 | 8 | 8640 | 6480 | 16.5 | 2.8 | 1.3 |
| 76 | 28 | 14 | 8.5 | 3332 | 2499 | 6.4 | 2.0 | 1.4 |
| 77 | 31 | 13 | 7 | 2821 | 2116 | 5.4 | 2.2 | 1.1 |
| Mean | 36 | 20.3 | 7.75 | 5664 | 4248 | 12.75 | 2.6 | 1.2 |

**Table A4.** Quantification of boulder size, volume, and estimated weight from all coastal bar boulders through transect 4 at Bahía San Luis Gonzaga. EWH = estimated wave height. The density of Baja Californian andesite at 2.3 gm/cm$^2$ is applied uniformly in order to calculate wave height for each boulder.

| Sample | Long Axis (cm) | Intermediate Axis (cm) | Short Axis (cm) | Volume (cm$^3$) | Adjust to 75% | Weight (kg) | EWH Nott [18] (m) | EWH Pepe [20] (m) |
|---|---|---|---|---|---|---|---|---|
| 1 | 27 | 20 | 10 | 5400 | 4050 | 10 | 1.9 | 1.6 |
| 2 | 70 | 39 | 30 | 81,900 | 61,425 | 157 | 4.9 | 4.8 |
| 3 | 39 | 20 | 12 | 9360 | 7020 | 18 | 2.7 | 1.9 |
| 7 | 63 | 39 | 20 | 49,140 | 36,855 | 94 | 4.4 | 3.2 |
| 8 | 75 | 38 | 35 | 99,750 | 74,813 | 191 | 5.3 | 5.6 |
| 9 | 25 | 14 | 12 | 4200 | 3150 | 8 | 1.8 | 1.9 |
| 10 | 38 | 29 | 19 | 20,938 | 15,704 | 40 | 2.7 | 3.0 |
| 12 | 29 | 23 | 17 | 11,339 | 8504 | 22 | 2.0 | 2.7 |
| 13 | 43 | 27 | 19 | 22,059 | 16,544 | 42 | 3.0 | 3.0 |
| 14 | 42 | 26 | 16 | 17,472 | 13,104 | 33 | 3.0 | 2.5 |
| 15 | 28 | 11 | 7 | 2156 | 1617 | 4 | 2.0 | 1.1 |
| 16 | 28 | 20 | 8 | 4480 | 3360 | 8.6 | 2.0 | 1.3 |
| 17 | 29 | 20 | 10 | 5800 | 4350 | 11 | 2.0 | 1.6 |
| 18 | 48 | 25 | 20 | 24,000 | 18,000 | 46 | 3.4 | 3.2 |
| 19 | 63 | 27 | 23 | 39,123 | 29,342 | 75 | 4.4 | 3.7 |
| 20 | 27 | 14 | 13 | 4914 | 3686 | 9 | 1.9 | 2.1 |
| 22 | 44 | 25 | 10 | 11,000 | 8250 | 21 | 3.1 | 1.6 |
| 25 | 40 | 34 | 15 | 20,400 | 15,300 | 39 | 2.8 | 2.4 |
| 26 | 32 | 23 | 5 | 3680 | 2760 | 7 | 2.3 | 0.8 |
| 27 | 61 | 25 | 19 | 28,975 | 21,731 | 55 | 4.3 | 3.0 |
| 29 | 38 | 27 | 13 | 13,338 | 10,004 | 26 | 2.7 | 2.1 |
| 31 | 40 | 16 | 15 | 9600 | 7200 | 18 | 2.8 | 2.4 |
| 32 | 50 | 34 | 25 | 42,500 | 31,875 | 81 | 3.5 | 4.0 |
| 34 | 35 | 19 | 18 | 11,970 | 8978 | 23 | 2.5 | 2.9 |
| 35 | 35 | 23 | 10 | 8050 | 6038 | 15 | 2.5 | 1.6 |
| 36 | 33 | 24 | 10 | 7920 | 5940 | 15 | 2.3 | 1.6 |
| 37 | 28 | 13 | 20 | 7280 | 5460 | 14 | 2.0 | 3.2 |
| 40 | 27 | 20 | 11 | 5940 | 4455 | 11 | 1.9 | 1.8 |
| 41 | 30 | 21 | 10 | 6300 | 4725 | 12 | 2.1 | 1.6 |
| 42 | 40 | 20 | 13 | 10,400 | 7800 | 20 | 2.8 | 2.1 |
| 45 | 40 | 38 | 10 | 15,200 | 11,400 | 29 | 2.8 | 1.6 |
| 48 | 27 | 17 | 10 | 4590 | 3443 | 9 | 1.9 | 1.6 |
| Mean | 40 | 24 | 15 | 19,037 | 14,278 | 36 | 2.8 | 2.4 |

**Table A5.** Quantification of boulder size, volume, and estimated weight from all coastal bar boulder through transect 5 at Bahía San Luis Gonzaga. The density of Baja Californian andesite at 2.3 gm/cm$^2$ is applied uniformly in order to calculate wave height for each boulder.

| Sample | Long Axis (cm) | Intermediate Axis (cm) | Short Axis (cm) | Volume (cm$^3$) | Adjust to 75% | Weight (kg) | EWH Nott [18] (m) | EWH Pepe [20] (m) |
|---|---|---|---|---|---|---|---|---|
| 1 | 57 | 40 | 38 | 86,640 | 64,980 | 166 | 4.0 | 6.1 |
| 2 | 38 | 26 | 9 | 8892 | 6669 | 17 | 2.7 | 1.4 |
| 3 | 35 | 32 | 4 | 4480 | 3360 | 8.6 | 2.5 | 0.6 |
| 7 | 40 | 23 | 10 | 9200 | 6900 | 17.6 | 2.8 | 1.6 |
| 8 | 27 | 20 | 9 | 4860 | 3645 | 9.3 | 1.9 | 1.4 |
| 10 | 32 | 23 | 10 | 7360 | 5520 | 14 | 2.3 | 1.6 |
| 18 | 26 | 24 | 14 | 8736 | 6552 | 17 | 1.8 | 2.2 |
| 19 | 25 | 12 | 10 | 3000 | 2250 | 5.7 | 1.8 | 1.6 |
| 20 | 36 | 27 | 14 | 13,608 | 10,206 | 26 | 2.5 | 2.2 |

**Table A5.** *Cont.*

| Sample | Long Axis (cm) | Intermediate Axis (cm) | Short Axis (cm) | Volume (cm$^3$) | Adjust to 75% | Weight (kg) | EWH Nott [18] (m) | EWH Pepe [20] (m) |
|---|---|---|---|---|---|---|---|---|
| 24 | 35 | 23 | 12 | 9660 | 7245 | 18.5 | 2.5 | 1.9 |
| 29 | 55 | 31 | 25 | 42,625 | 31,969 | 82 | 3.9 | 4.0 |
| 30 | 44 | 40 | 23 | 4480 | 3360 | 8.6 | 3.1 | 3.7 |
| 33 | 37 | 33 | 26 | 31,746 | 23,910 | 61 | 2.6 | 4.1 |
| 37 | 25 | 18 | 17 | 7650 | 5738 | 14.6 | 1.8 | 2.7 |
| 38 | 38 | 23 | 8 | 6992 | 5244 | 13 | 2.7 | 1.3 |
| 44 | 30 | 19 | 12 | 6840 | 5130 | 13 | 2.1 | 1.9 |
| 46 | 79 | 43 | 21 | 71,337 | 53,503 | 136 | 5.6 | 3.3 |
| 47 | 35 | 17 | 12 | 7140 | 5355 | 14 | 2.5 | 1.9 |
| 56 | 36 | 20 | 12 | 8640 | 6480 | 17 | 2.5 | 1.9 |
| 101 | 42.5 | 28 | 19 | 22,610 | 16,958 | 43 | 3.0 | 3.0 |
| 102 | 39 | 23 | 13 | 11,661 | 8746 | 22 | 2.7 | 2.1 |
| 103 | 42 | 24 | 6 | 6048 | 4536 | 11.5 | 3.0 | 1.0 |
| 104 | 26 | 21 | 10 | 5460 | 4095 | 10 | 1.8 | 1.6 |
| 106 | 36 | 25 | 11 | 9900 | 7425 | 19 | 2.5 | 1.8 |
| 107 | 45 | 18 | 14 | 11,340 | 8505 | 22 | 3.2 | 2.2 |
| 108 | 34 | 24 | 7 | 5712 | 4284 | 11 | 2.4 | 1.1 |
| 115 | 28 | 26 | 12 | 8736 | 6552 | 17 | 2.0 | 1.9 |
| 119 | 40 | 30 | 12 | 14,400 | 10,800 | 27.5 | 2.8 | 1.9 |
| 123 | 25 | 15 | 15 | 5625 | 4219 | 11 | 1.8 | 2.4 |
| Mean | 37.5 | 25 | 14 | 15,358 | 11,519 | 29.4 | 2.6 | 2.2 |

**Table A6.** Quantification of boulder size, volume, and estimated weight from all coastal bar boulders through transect 6 at Bahía San Luis Gonzaga. EWH = estimated wave height. The density of Baja Californian andesite at 2.3 gm/cm$^2$ is applied uniformly in order to calculate wave height for each boulder.

| Sample | Long Axis (cm) | Intermediate Axis (cm) | Short Axis (cm) | Volume (cm$^3$) | Adjust to 75% | Weight (kg) | EWH Nott [18] (m) | EWH Pepe [20] (m) |
|---|---|---|---|---|---|---|---|---|
| 1 | 33 | 22 | 22 | 15,972 | 11,979 | 30.5 | 2.3 | 3.5 |
| 2 | 37 | 30 | 12 | 13,320 | 9990 | 17 | 2.6 | 1.9 |
| 3 | 39 | 33 | 5 | 6435 | 4826 | 8.6 | 2.7 | 0.8 |
| 15 | 40 | 23 | 14 | 12,880 | 9660 | 24.6 | 2.8 | 2.2 |
| 16 | 39 | 30 | 25 | 29,250 | 21,938 | 56 | 2.7 | 4.0 |
| 20 | 29 | 26 | 6 | 4524 | 3393 | 8.6 | 2.0 | 1.0 |
| 26 | 29 | 19 | 11 | 6061 | 4546 | 12 | 2.0 | 1.8 |
| 30 | 80 | 54 | 23 | 99,360 | 74,520 | 190 | 5.6 | 3.7 |
| 41 | 26 | 21 | 6 | 3276 | 2457 | 6 | 1.8 | 1.0 |
| 69 | 31 | 26 | 7 | 5642 | 4232 | 11 | 2.2 | 1.1 |
| 73 | 44 | 16 | 14 | 9856 | 7392 | 19 | 3.1 | 2.2 |
| 80 | 27 | 14 | 8 | 3024 | 2268 | 6 | 1.9 | 1.3 |
| 83 | 26 | 18 | 13 | 6084 | 4563 | 12 | 1.8 | 2.1 |
| 93 | 35 | 15 | 7 | 3675 | 2756 | 7 | 2.5 | 1.1 |
| 94 | 32 | 24 | 23 | 17,664 | 13,248 | 34 | 2.3 | 3.7 |
| 96 | 37 | 25 | 11 | 10,175 | 7631 | 20 | 2.6 | 1.8 |
| 97 | 26 | 16 | 13 | 5408 | 4056 | 10 | 1.8 | 2.1 |
| 101 | 38 | 16 | 12 | 7296 | 5472 | 14 | 2.7 | 1.9 |
| 102 | 26 | 13 | 11 | 3718 | 2789 | 7 | 1.8 | 1.8 |
| 103 | 34 | 19 | 9 | 5814 | 4361 | 11 | 2.4 | 1.4 |
| Average | 35 | 23 | 12.6 | 13,472 | 10,104 | 25 | 2.5 | 2.0 |

**Table A7.** Quantification of boulder size, volume, and estimated weight from selected coastal bar samples through transect 7 at Bahía San Luis Gonzaga. EWH = estimated wave height. The density of Baja Californian andesite at 2.3 gm/cm$^2$ is applied uniformly in order to calculate wave height for each boulder.

| Sample | Long Axis (cm) | Intermediate Axis (cm) | Short Axis (cm) | Volume (cm$^3$) | Adjust to 75% | Weight (kg) | EWH Nott [18] (m) | EWH Pepe [20] (m) |
|---|---|---|---|---|---|---|---|---|
| 5 | 53 | 49 | 32 | 83,104 | 62,328 | 159 | 3.7 | 5.1 |
| 7 | 58 | 41 | 20 | 47,560 | 35,670 | 91 | 4.1 | 3.2 |
| 12 | 26 | 15 | 12 | 5760 | 4320 | 11 | 1.8 | 1.9 |
| 15 | 34 | 22 | 15 | 11,220 | 8415 | 22 | 2.4 | 2.4 |
| 17 | 38 | 26 | 15 | 14,820 | 11,115 | 28 | 2.7 | 2.4 |
| 21 | 55.5 | 53 | 41 | 120,602 | 90,451 | 231 | 3.9 | 6.5 |
| 24 | 44 | 25 | 22 | 24,200 | 18,150 | 46 | 3.1 | 3.5 |
| 29 | 30 | 29 | 24 | 20,880 | 15,660 | 40 | 2.1 | 3.8 |
| 31 | 36 | 20 | 20 | 14,400 | 10,800 | 28 | 2.5 | 3.2 |
| 34 | 39 | 27 | 22 | 23,166 | 17,375 | 44 | 2.7 | 3.5 |
| 37 | 34 | 30 | 14 | 14,280 | 10,710 | 27 | 2.4 | 2.2 |
| 41 | 28 | 15 | 10 | 4200 | 3150 | 8 | 2.0 | 1.6 |
| 47 | 27 | 23 | 19 | 11,799 | 8849 | 23 | 1.9 | 3.0 |
| 50 | 44 | 20 | 14 | 12,320 | 9240 | 24 | 3.1 | 2.2 |
| 55 | 50 | 48 | 35 | 84,000 | 63,000 | 161 | 3.5 | 5.6 |
| 64 | 29 | 25 | 24 | 17,400 | 13,050 | 33 | 2.0 | 3.8 |
| 67 | 27 | 24 | 5 | 3240 | 2430 | 6 | 1.9 | 0.8 |
| 71 | 30 | 27 | 14 | 11,340 | 8505 | 22 | 2.1 | 2.2 |
| 73 | 26 | 22 | 14 | 8008 | 6006 | 15 | 1.8 | 2.2 |
| 79 | 25 | 16 | 12 | 4800 | 3600 | 9 | 1.8 | 1.9 |
| 81 | 52 | 29 | 13 | 19,604 | 14,703 | 38 | 3.7 | 2.1 |
| 83 | 27.5 | 19 | 12 | 6270 | 4703 | 12 | 1.9 | 1.9 |
| 88 | 30 | 26 | 14 | 10,920 | 8190 | 21 | 2.1 | 2.2 |
| 94 | 43 | 43 | 25 | 46,225 | 34,669 | 88 | 3.0 | 4.0 |
| 97 | 44 | 30 | 20 | 26,400 | 19,800 | 51 | 3.1 | 3.2 |
| 100 | 30 | 21 | 13 | 8190 | 6143 | 16 | 2.1 | 2.1 |
| 103 | 35 | 20 | 10 | 7000 | 5250 | 13 | 2.5 | 1.6 |
| 105 | 26 | 23 | 16 | 9568 | 7176 | 18 | 1.8 | 2.5 |
| 107 | 84 | 38 | 20 | 63,840 | 47,880 | 122 | 5.9 | 3.2 |
| 112 | 34 | 17 | 15 | 8670 | 6503 | 17 | 2.4 | 2.4 |
| Mean | 38 | 27 | 18 | 24,793 | 18,595 | 47 | 2.7 | 2.9 |

# References


1. Romero-Vadillo, E.; Zaystev, O.; Morales-Pérez, R. Tropical cyclone statistics in the northeastern Pacific. *Atmósfera* **2007**, *20*, 197–213.
2. Muriá-Vila, D.; Jaimes, M.Á.; Pozos-Estrada, A.; López, A.; Reinoso, E.; Chávez, M.M.; Peña, F.; Sánchez-Sesma, J.; López, O. Effects of hurricane Odile on the infrastructure of Baja California Sur, Mexico. *Nat. Hazards* **2018**, *9*, 963–981. [CrossRef]
3. Gross, M.; Magar, V. Wind-induced currents in the Gulf of California from extreme events and their impact on tidal energy devices. *J. Mar. Sci. Eng.* **2020**, *8*, 80. [CrossRef]
4. Merrifield, M.A.; Winant, C.D. Shelf-circulation in the Gulf of California: A description of the variability. *J. Geophys. Res.* **1989**, *94*, 133–160. [CrossRef]
5. May, S.M.; Engel, M.; Brill, D.; Squire, P.; Scheffers, A.; Kelletat, D. Coastal hazards from tropical cyclones and extratropical winter storms based on Holocene storm chronologies. In *Coastal Hazards*; Finkl, C.W., Ed.; Coastal Research Library: Cham, Switzerland, 2013; Volume 6, pp. 557–585.
6. Switzer, A.D.; Burston, J.M. Competing mechanisms for boulder deposition on the southeast Australian coast. *Geomorpholgy* **2010**, *114*, 42–54. [CrossRef]

7.   Buchanan, D.H.; Naylor, L.A.; Hurst, M.D.; Stephenson, W.J. Erosion of rocky shore platforms by block detachment from layered stratigraphy. *Earth Surf. Process. Landf.* **2019**, *45*, 1028–1037. [CrossRef]

8.   Suursaar, Ü; Alari, V.; Tõnisson, H. Multi-scale analysis of wave conditions and coastal changes in the northeastern Baltic Sea. *J. Coast. Res.* **2014**, *70*, 223–228. [CrossRef]

9.   Johnson, M.E.; Ledesma-Vázquez, J.; Guardado-France, R. Coastal geomorphology of a Holocene hurricane deposit on a Pleistocene marine terrace from Isla Carmen (Baja California Sur, Mexico). *J. Mar. Sci. Eng.* **2018**, *6*, 108. [CrossRef]

10.  Johnson, M.E.; Guardado-France, R.; Johnsob, E.M.; Ledesma-Vázquez, J. Geomorphology of a Holocene Hurricane deposit eroded from rhyolite sea cliffs on Ensenada Almeja (Baja California Sur, Mexico). *J. Mar. Sci. Eng.* **2019**, *7*, 193. [CrossRef]

11.  Johnson, M.E.; Johnson, E.M.; Guardado-France, R.; Ledesma-Vázquews, J. Holocene hurricane deposits eroded as coastal barriers from andesite sea cliffs at Puerto Escondido (Baja California Sur, Mexico). *J. Mar. Sci. Eng.* **2020**, *8*, 75. [CrossRef]

12.  Backus, D.H.; Johnson, M.E.; Ledesma-Vazquez, J. Peninsular and island rocky shores in the Gulf of California. In *Atlas of Coastal Ecosystems in the Western Gulf of California*; Johnson, M.E., Ledesma-Vazquez, J., Eds.; University Arizona Press: Tucson, AZ, USA, 2009; pp. 11–27. ISBN 978-0-8165-2530-0.

13.  Gastil, R.G.; Phillips, R.P.; Allison, E.C. *Reconnaissance Geological Map of the State of Baja California*; Geological Society America; Map Sheets a, b, and c; Geological Society of America: Boulder, CO, USA, 1971.

14.  Wentworth, C.K. A scale of grade and class terms for clastic sediments. *J. Geol.* **1922**, *27*, 377–392. [CrossRef]

15.  Sneed, E.D.; Folk, R.L. Pebbles in the lower Colorado River of Texas: A study in particle morphogenesis. *J. Geol.* **1958**, *66*, 114–150. [CrossRef]

16.  Yeh, F.-H.; Huang, C.-J.; Han, J.-Y.; Ge, L. Modeling slope topography using unmanned aerial vehicle image technique. *MATEC Web Conf.* **2018**, *147*, 07002. [CrossRef]

17.  Casella, E.; Drechsel, J.; Winbter, C.; Benninghoff, M.; Rovere, A. Accuracy of sand beach topography surveying by drones and photogrammetry. *Geo-Mar. Lett.* **2020**, *40*, 255–268. [CrossRef]

18.  Nott, J. Waves, coastal bolder deposits and the importance of pre-transport setting. *Earth Planet. Sci. Lett.* **2003**, *210*, 269–276. [CrossRef]

19.  Nandasena, N.A.K.; Paris, R.; Tanaka, N. Reassessment of hydrodynamic equations: Minimum flow velocity to initiate boulder transport by high energy events (storms, tsunamis). *Mar. Geol.* **2011**, *281*, 70–84. [CrossRef]

20.  Pepe, F.; Corradino, M.; Parrino, N.; Besio, G.; Presti, V.L.; Renda, P.; Calcagnile, L.; Quarta, G.; Sulli, A.; Antonioli, F. Boulder coastal deposits at Favignana Island rocky coast (Sicily, Italy): Litho-structural and hydrodynamic control. *Geomorphology* **2018**, *303*, 191–209. [CrossRef]

21.  Rebman, J.P.; Roberts, N.C. *Baja California Plant Field Guide*, 3rd ed.; Sun Belt Publications: San Diego, CA, USA, 2012; p. 451. ISBN 978-0-916251-18-5.

22.  Brusca, R.C. *Common Intertidal Invertebrates of the Gulf of California*, 2nd ed.; University of Arizona Press: Tucson, AZ, USA, 1980; p. 513. ISBN 0-8165-0682-5.

23.  Nott, J.; Haig, J.; Neil, H.; Gillierson, D. Greater frequency of landfalling tropical cyclones at centennial compared to seasonal and decadal scales. *Earth Planet. Sci. Let.* **2007**, *255*, 367–372. [CrossRef]

24.  Avila, L. The 2015 Eastern North Pacific Hurricane Season: A very active year. *Weatherwise* **2016**, *69*, 36–42. [CrossRef]

25.  Kossin, J.P.; Knapp, K.R.; Olander, T.L.; Velden, C.S. Global increase in major tropical cyclone exceedance probability over the past four decades. *Proc. Natl. Acad. Sci. USA* **2020**, *117*, 11975–11980. [CrossRef]

26.  Ávila, S.P.; Johnson, M.E.; Rebelo, A.C.; Baptista, L.; Melo, C.S. Comparison of modern and Pleistocene (MIS 5e) coastal Boulder deposits from Santa Maria Island (Azores Archipelago, NE Atlantic Ocean). *J. Mar. Sci. Eng.* **2020**, *8*, 386. [CrossRef]

Journal of
*Marine Science
and Engineering*

*Article*

# Holocene Hurricane Deposits Eroded as Coastal Barriers from Andesite Sea Cliffs at Puerto Escondido (Baja California Sur, Mexico)

**Markes E. Johnson** [1,*], **Erlend M. Johnson** [2], **Rigoberto Guardado-France** [3] and **Jorge Ledesma-Vázquez** [3]

[1]  Geosciences Department, Williams College, Williamstown, MA 01267, USA
[2]  Anthropology Department, Tulane University, New Orleans, LA 70018, USA; erland.johnson@gmail.com
[3]  Facultad de Ciencias Marinas, Universidad Autónoma de Baja California, Ensenada 22800, Baja California, Mexico; Rigoberto@uabc.edu.mx (R.G.-F.); Ledesma@uabaac.edu.mx (J.L.-V.)
*  Correspondence: mjohnson@williams.edu; Tel.: +1-413-597-2329

Received: 1 November 2019; Accepted: 22 January 2020; Published: 24 January 2020

**Abstract:** Previous studies on the role of hurricanes in Mexico's Gulf of California examined coastal boulder deposits (CBDs) eroded from limestone and rhyolite sea cliffs. Sedimentary and volcanic in origin, these lithotypes are less extensively expressed as rocky shores than others in the overall distribution of gulf shores. Andesite that accumulated as serial volcanic flows during the Miocene constitutes by far the region's most pervasive rocky shores. Here, we define a subgroup of structures called barrier boulder deposits (BBDs) that close off lagoons as a result of lateral transport from adjacent rocky shores subject to recurrent storm erosion. Hidden Harbor (*Puerto Escondido*) is the most famous natural harbor in all of Baja California. Accessed from a single narrow entrance, it is commodious in size (2.3 km$^2$) and fully sheltered by outer andesite hills linked by two natural barriers. The average weight of embedded boulders in a succession of six samples tallied over a combined distance of 710 m ranges between 74 and 197 kg calculated on the basis of boulder volume and the specific gravity of andesite. A mathematical formula is utilized to estimate the wave height necessary to transport large boulders from their source. Average wave height interpreted by this method varies between 4.1 and 4.6 m. Input from fossil deposits and physical geology related to fault trends is applied to reconstruct coastal evolution from a more open coastal scenario during the Late Pleistocene 125,000 years ago to lagoon closure in Holocene time.

**Keywords:** barrier boulder deposits; hurricane storm surge; hydrodynamic equation; Gulf of California (Mexico)

---

## 1. Introduction

Based on a coastal survey using satellite imagery [1], volcanic flows of Miocene age that accrued as andesite were found to account for more than 700 km of peninsular and island shores in the western Gulf of California. By far, andesite is the most common rock type, accounting for 24% of all shores including sand beaches. Given the dominant occurrence of these rocks, it is pertinent to ask how it responds to forces of physical erosion. This contribution is the third in a series to examine rocky shores in the context large-scale boulder deposits attributed to storms of hurricane intensity that impacted the peninsular inner shores of Mexico's Baja California. The strength and behavior of recent storms, such as Hurricane Odile in 2014, follow a consistent pattern that allows predictions to be tested as to the specific vulnerability of different rock types. Previous work focused on a coastal boulder deposit (CBD) with metric-ton blocks of Pliocene limestone torn from the outer margin of a 12-m marine terrace on Isla del Carmen in the Gulf of California [2]. A subsequent study examined similar-size boulders pried from a rhyolite coast not far to the north at Ensenada Almeja [3]. Limestone rocky shores amount to only 7.5% of shores in the western Gulf of California and rhyolite is so uncommon, it was not part of our original satellite reconnaissance [1].

Here, we consider the natural setting at Puerto Escondido (Spanish for Hidden Harbor), which has a restricted entrance but opens to a large lagoon otherwise entirely surrounded by andesite foothills related to the rugged Sierra de la Giganta. The inner lagoon is large enough to accommodate a small armada and the working harbor has been modernized to accommodate anchorage for visiting yachts and sailboats as well as larger vessels that call at the main wharf just inside the entrance. Puerto Escondido is renowned for the description by Steinbeck and Ricketts [4] during their epic voyage to the Gulf of California aboard the *Western Flyer* in 1940. Marine biologist, Ed Ricketts, who in 1939 published a ground-breaking treatise on the intertidal relationships of marine invertebrates along the Pacific shores of the United States [5], planned the expedition to expand his observations to the biologically rich but then poorly studied Sea of Cortez. His friend, author John Steinbeck, called the Hidden Harbor a place of magic and wrote: "If one wished to design a secret personal bay, one would probably build something very like this little harbor."

The goal of this paper is to apply geological and geomorphological insights to explain why Puerto Escondido is so extraordinary as a natural harbor. Paleontological data, as well as the location of critical fault lines, are used to show how the coast was more open to marine circulation during the last interglacial epoch in the Late Pleistocene. The emplacement of two major barriers fixed among outer hills is the primary focus of analysis looking at boulder shapes and their variation in size and calculated weight. Estimation of wave heights necessary to transport large boulders serves as a proxy to gauge recurrent storm intensity. Lastly, geomorphologic modeling provides a means to consider the degree to which rocky-shore retreat has occurred over Holocene time in a consistently subtropical setting and the scale of erosion necessary to provide the raw materials for barrier construction.

## 2. Geographical and Geological Setting

Situated between the Mexican mainland and the Baja California peninsula, the Gulf of California is a marginal sea with a semi-enclosed area amounting to 210,000 km$^2$ arrayed along a NW–SE axis stretching for 1100 km (Figure 1a). Central basins within the gulf are semi-oceanic in depth, exceeding 3200 m. The southern opening to the Pacific Ocean is 180 km wide and allows for a range of oceanographic phenomena [6] that in turn stimulates seasonal upwelling and nutrient fertilization linked to a high degree of biological productivity and species diversity [7]. More than a dozen tropical storms typically form off the cost of Acapulco at approximately 15° N latitude during the annual hurricane season, but most turn outward to the northwest before reaching the southern tip of the Baja California peninsula at 23° N Latitude [8]. Major storms are known to enter the Gulf of California—most recently, Hurricane Odile in September 2015 [9] and Hurricane Lorena in September 2019.

Puerto Escondido is located 24 km south from the town of Loreto in Baja California Sur on the Gulf of California (Figure 1b, locality 1). A small outer harbor is linked to a huge inner harbor by a narrow entrance on the south side. Viewed from hills on the inland western side (Figure 2), the harbor is notable for a pair of distinct barriers that form robust sea walls anchored to an intermediate islet. At their opposite ends, the pair of barriers are linked to the bedrock on the peninsula mainland to the north and a large island to the south that also guards the entrance to the inner harbor. In concert, the combination of natural barriers and fixed bedrock effectively seals off the lagoon from outside disturbances in the open Gulf of California.

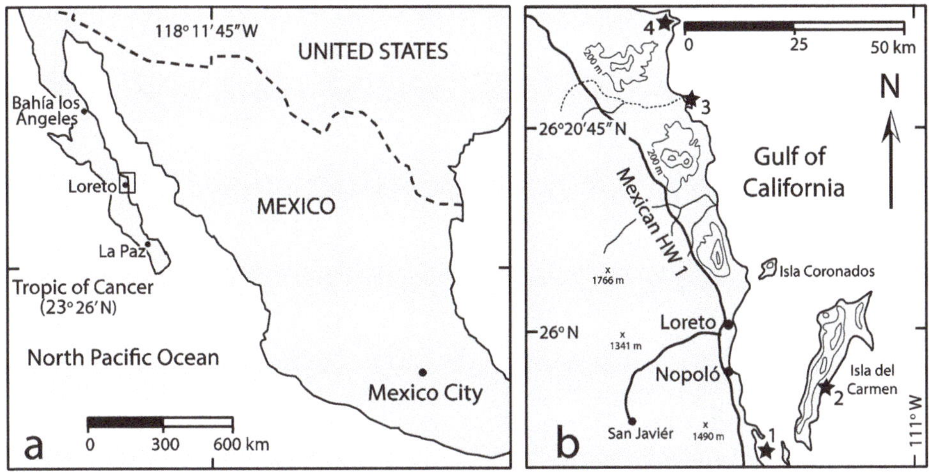



**Figure 1.** Locality maps showing Mexico's Baja California peninsula and Gulf of California; (**a**) Mexico and border area with the Unite States, denoting key villages or cities with inset box marking the study region around the town of Loreto; (**b**) Region around Loreto in Baja California Sur, marking coastal boulder deposits (*) at localities 1 to 4.

**Figure 2.** View east over the inner harbor at Puerto Escondido with Isla del Carmen on the horizon.

From a geological perspective, the history of faulting in western Mexico is intimately related to the origins of the Gulf of California. Tectonic separation of the Baja California peninsula from the mainland occurred due to crustal extension between 13 and 3.5 million years ago, with N–S trending faults related to Basin and Range development in western North America. Thereafter, a change in tectonic regime led to transtensional faulting with the transfer of the peninsula to the Pacific Tectonic Plate and its ongoing migration to the NW [7]. During the earlier phase, faults were oriented mainly N–S and the peninsular coast underwent major uplift west of the Loreto rift segment between 5.6 and 3.2 million years ago, amounting to 100s of meters in the Sierra de la Gigante [10]. Many of the 40 named islands in the gulf conform to fault blocks subjected to uplift as structural horsts [11]. Subsequent strike-slip faulting is oriented NW–SE perpendicular to a series of step-like spreading centers located within deep basins through the Gulf of California. A major clue as to the tectonic history of the study area around Puerto Escondido is the placement of faults registered in the landscape.

## 3. Materials and Methods

### 3.1. Data Collection

Data on topography, basin size, watershed boundaries and fault orientations are derived from a portion of the Mexican federal government map for the Juncalito quadrangle (G12C19). Drawn at a scale of 1:50,000, the greater map was issued by the Instituto Nacional de Estadistica Geografia e Informatica in 1982. Contour intervals from the government map were scanned and traced to yield a project map retaining an accuracy at 20 m intervals. The 1982 version of the map was updated to show the main details of improved harbor infrastructure since that time.

Puerto Escondido was visited on 24 and 25 April 2019, when the field data for this study were collected based foremost on a sample of 100 boulders divided equally among four transects along the upper tide line of the northern Coastal Barrier Deposit (CBD) and another 50 boulders along the southern CBD. The boundary is denoted by a prominent color due to marine algae. The definition for a boulder adapted in this exercise is that of Wentworth (1922) for an erosional clast equal or greater than 256 mm in diameter [12]. There exists no proposed upper limit in size for this category.

Collection of data on boulder size followed procedures graphically codified in Figure 3. A Brunton compass and meter tape were used to lay out transects in 50 m segments. Consistent with previous studies [1,2], the largest 25 boulders were measured manually in each transect with boulder centers spaced from 1 to 1.5 m apart. Each boulder required three measurements along principle axes (long a, intermediate b and short c). Triangular plots were employed to demonstrate variations in boulder shape, following the practice of Sneed and Folk (1958) for river pebbles [13]. Data regarding the maximum and intermediate lengths perpendicular to one another from individual boulders were plotted in bar graphs to show potential shifts in size from one transect to the next. A representative cobble of andesite was collected from the northern barrier for laboratory treatment at Williams College, where it was weighed, and its volume determined as a function of equal displacement when submerged in a beaker of water. Prior to immersion, the rock was water-proofed by spraying it with Thompson's Water Seal TM (The Thompson's Co, Cleveland, OH, USA).

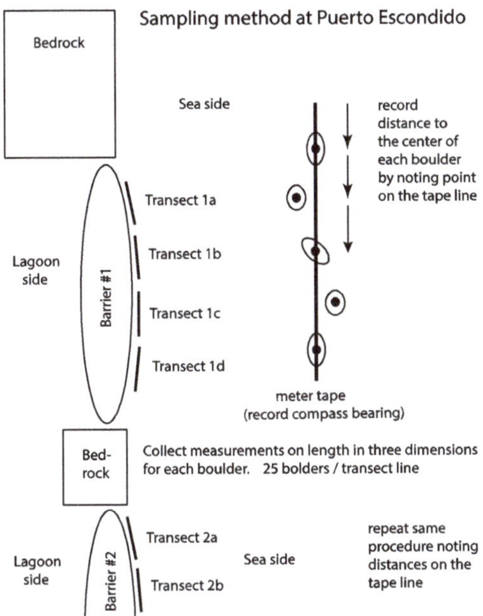


**Figure 3.** Schematic portrayal of the sampling method applied at Puerto Escondido (not to scale).

## 3.2. Hydraulic Model

With determination of specific gravity based on laboratory testing for volume and weight, a hydraulic model may be applied to predict the energy needed to transport larger andesite blocks from a rocky shoreline to a barrier deposit as a function of wave impact. Andesite is a volcanic rock that forms from surface flows with variable thicknesses and a propensity to vertical fractures. These factors control the size and general shape of blocks loosened in the cliff face. Herein, the formula used to estimate the magnitude of storm waves applied to joint-bounded boulders is taken from equation 36 in the work of Nott [14]:

$$Hs = \frac{(Ps - Pw/Pw)\,a}{C_1}$$

where $Hs$ = height of the storm wave at breaking point; $Ps$ = density of the boulder (tons/m$^3$ or g/cm$^3$) $Pw$ = density of water at 1.02 g/mL; a = length of boulder on long axis in cm; and $C_1$ = lift coefficient (=0.178).

## 4. Results

### 4.1. Topographic Base Map

The base map adapted for use in this project treats an area of 25 km$^2$ (Figure 4). A small outer harbor open to the south occupies an area of 0.5 km$^2$. To one side of the outer harbor, the mouth (*La Bocana*) forms a 50-m wide entrance to a much larger inner harbor covering 2.3 km$^2$. The narrow connection between outer and inner harbors admits tidal flux but resists severe weather arriving from all directions. Hills surrounding the inner harbor inland to the west exhibit lower topography with elevations ranging between 100 and 160 m above sea level. The outer eastern edge of the harbor complex is formed by a linear front stretching 4 km from NW to SE consisting of two hills (Cerro El Chino and Cerro La Enfermería) connected to an un-named islet by barriers #1 and #2 (Figure 4). The islet rises to an elevation exceeding 80 m above sea level, whereas bedrock on the neighboring hills reaches 120 and 180 m, respectively. Formed by andesite pebbles, cobbles, and boulders, the two natural breakwaters are the most vulnerable spots in the outer defense of the main harbor. The longer northern barrier extends for 250 m, whereas the shorter southern barrier is 140 m in length. On average, barrier width amounts to 30 m, with a mid-line 2.75 m above mean sea level. On close inspection, the inner west-facing edges of the barriers drop off abruptly into the enclosed lagoon. The outer east-facing margins are ramp-like in configuration extending at a low angle into the water.

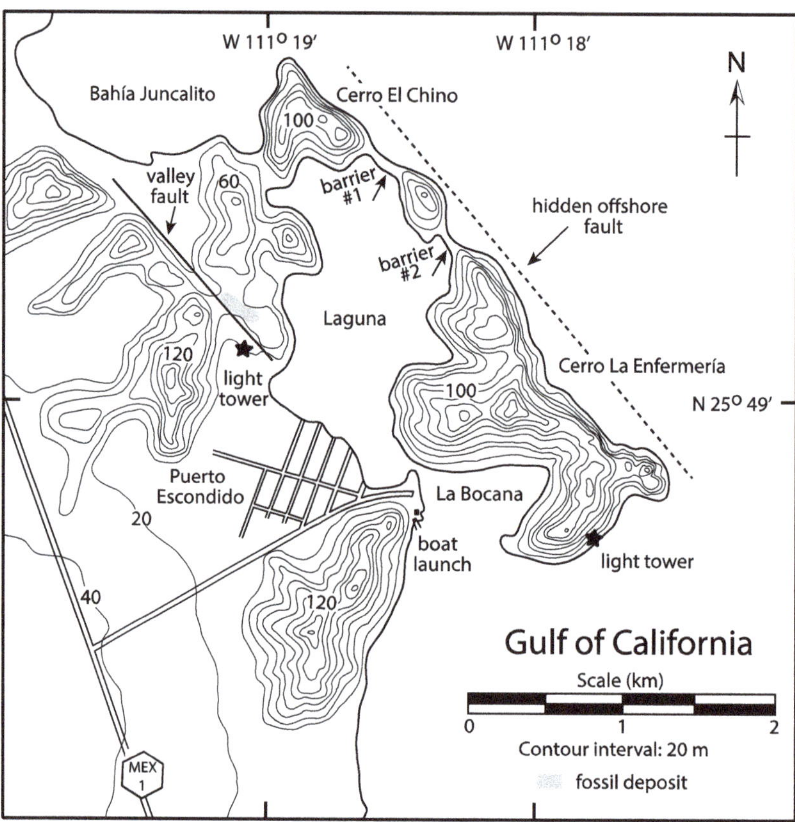


**Figure 4.** Topographic map showing the hills surrounding Puerto Escondido and other key features including a Pleistocene fossil deposit and faults.

### 4.2. Paleontological Data from the Western Hills

Described here for the first time, a key fossil deposit occurs in the western hills above the inner harbor at Puerto Escondido. Approximately one hectare in area (Figure 5a), the deposit is comparable to Pleistocene shell drapes found on 12-m marine terraces along of the peninsular gulf coast [15]. Here, the shell drape sits 45 m above present sea level covering the topographic saddle between hills separating the harbor lagoon on one side and Bahía Juncalito to the north (Figure 4). Loose shells in the deposit correlate with the last interglacial epoch 125,000 years ago, when sea level worldwide was 6 m higher than today based on comparisons with marine deposits from islands regarded as tectonically stable [16,17]. A precise radiometric date is not possible, because datable *Porites* corals are not found at this locality. For the most part, the Puerto Escondido drape consists of white-bleached and abundant shells from the clam *Chione californiensis*—all of which are disarticulated as separate valves. Rare, but easy to spot within the mix is a small oyster (*Ostrea fischeri*), also known to encrust rocks in an intertidal environment. In a single sample covering 16 dm$^2$, approximately 75 valves of the dominant *Chione* clam litter the surface (Figure 5b). Tested laterally, the deposit is rarely more than 15 cm in thickness.

**Figure 5.** Upper Pleistocene shell drape in the western hills around Puerto Escondido: (**a**) View from an elevation of 45 m above sea level looking west toward the Sierra de la Gigante in the background; (**b**) Close-up view of the shell drape dominated by disarticulated vales of the Mollusk bivalve, *Chione californiensis*. Pocket knife for scale is 9 cm in length.

The tally features whole (unbroken) valves, although fragmented shells also are present. From the sample, there is scarcely any evidence for other species of marine mollusks. The operculum of a marine gastropod (*Turbo fluctuosus*) is small and easy to overlook. That particular species is diagnostic for a herbiverous gastropod typically found living in an intertidal setting, today. Another shell hidden in the array is a predatory gastropod (*Murex elenensis*), also found today living in the intertidal zone where it feeds on other mollusks and barnacles. The *Turbo* utilizes a hard mouthpiece called a radula like a file to scrape off marine algae critical to its diet, whereas the *Murex* uses a similar mouth device to bore a hole through the shells of its prey to gain access for feeding. Larger shells are widely scattered around the deposit but few in number. They include the turkey shell (*Cardita megastrophica*), bittersweet shell (*Glycymeris maculate*), cholate shell (*Megapitaria squalida*), cockle shell (*Trachycardium panamense*), and rock oyster (*Spondylus calcifer*). The overall species list requires some effort to assemble, because these larger shells represent a clear minority within the shell assemblage. All are represented by extant species living in the Gulf of California today [18]. The fossil shell drape is observed to occur off to one side of a fault that extends southward through a narrow valley from Bahía Juncalito. Considering that shell drapes of this reputed age from marine terraces elsewhere typically occur 12 m above sea level [15], tectonic uplift of the hills around Puerto Escondido represents a local anomaly some 32 m above normal. Proximity to the master Loreto fault and massive uplift of the nearby Sierra de la Gigante [10] account for this discrepancy.

The original work by Steinbeck and Ricketts [4] includes a detailed accounting of marine life both within the inner harbor and the outer harbor. These local biological data make a useful contrast with the paleontological data. The Brown Cucumber (*Isostichopus fusca*), a holothurian typically 15 cm in length, was found to occupy the inner harbor in a population numbering in the hundreds. Sand flats on the west side of the inner harbor appeared to be sterile. In contrast, one of the richest collecting stations of the entire expedition was reported from the outer harbor, where tidal currents were strongest at the entrance. A diverse biota was recorded to include sponges, tunicates, chitons, limpets, bivalves, snails, hermit crabs, as well as numerous species of sea cucumbers and starfish.

### 4.3. Sample Density Calculation

Specific gravity is a precise physical property that compares the density of an object to that of one cubic centimeter of water. One of the diagnostic characteristics of all naturally occurring minerals is defined by a known specific gravity. Because the composition of igneous rocks is variable, depending on the ratio of component minerals, the specific gravity of a particular class of igneous rocks like granite, rhyolite, or andesite will vary from region to region. The andesite sample from Puerto Escondido collected for laboratory analysis was a single cobble determined to weigh 575 g. After treatment to make the sample water-tight, the cobble was submerged in a wide-mouth, graduated beaker partially filled with distilled water. The volume of water displaced through this operation amounted to 225 mL. Dividing mass by volume yielded a density of 2.55 for the andesite sample, which means it was found to be 2.55 times as dense as water. The laboratory result was subsequently applied uniformly to all further calculations using the formula cited in the methods section, above. For comparison, the specific gravity of limestone from our earlier study at nearby Isla del Carmen (Figure 1b, locality 2) was determined to be 1.86 [2] and the banded rhyolite from our study at Ensenada Almeja (Figure 1b, locality 3) was determined to be 2.16 [3].

### 4.4. Placement of Transects and Analysis of Boulder Shapes

The modern inter-tidal zone is visible on the exposed outer margin of the two barriers by boulders tinted light green in color with the growth of filamentous algae. Contrast with the clean supratidal zone is clearly observed looking to the NW on barrier #1 with the sea cliffs of Cerro El Chino in the background (Figure 6a). All transects in this study were set along this boundary at the top of the intertidal zone. In the opposite direction, the view shows placement of the meter tape with the un-named islet in the distance to the SE (Figure 6b).

It is worthy of note that both Cerro El Chino and the un-named islet exhibit steep sea cliffs where erosion resulted in significant coastal retreat. The effect is apparent in the asymmetry of the outer hills and related islet as registered on the topographic map (Figure 4). A view from the middle of barrier #1 looking SW shows a characteristic mixture of pebbles, cobbles, and boulders entrained in the natural breakwater (Figure 6b). Looking eastward (Figure 6c), the meter tape is extended in front of the figure and the larger boulders in the field of view fall along the boundary between the darker upper inter-tidal zone and the normal supratidal zone. Much the same mixture of eroded cobbles and boulders is visible in the shallow water behind the figure. The larger boulders are typically 1 m in maximum diameter.

Raw data on boulder size in three dimensions collected from four consecutive transects each—50 m in length—are available in Tables 1–4. Data points representing individual boulders grouped by transect are plotted on a set of Sneed-Folk triangular diagrams (Figure 7a–d), showing the actual variation in shapes. Those points clustered nearest to the core of the diagrams are most faithful to an average value with somewhat equidimensional axes in three directions. Only very seldom do points for these boulders appear in the upper-most triangle, which signifies a cube-shaped endpoint. The vast majority falls within the central part of the two tiers beneath the top triangle. However, the overall trend among those points grouped from different transects trace a similar pattern angled toward the lower right corner of the diagrams. No points are plotted in the lower left tier in any of the diagrams, but a few occur in the lower right tier most notably in Figure 7b,c. The repetitive pattern in these plots indicates a tendency toward boulders that are oblong in shape. Although the trends in shape are similar among the four samples from barrier #1, the plots have no bearing on variations in boulder size.

**Figure 6.** Boulder deposits from barrier #1; (**a**) View from the north end of the boulder deposit with the eroded cliff face of Cerro El Chino in the background; (**b**) View looking toward the south end with the eroded cliff face of an un-named islet in the distance; (**c**) View east from the center of the barrier. In each view, the anchor position of the meter tape is at the upper tide line marked by black arrows.

**Table 1.** Quantification of boulder size, volume and estimated weight from coastal bar samples through Transect 1a at the east end of Cerro El Chino (Puerto Escondido). The laboratory result for density of andesite at 2.55 gm/cm$^3$ is applied uniformly in order to calculate wave height for each boulder.

| Sample | Distance to Next (cm) | Long Axis (cm) | Intermediate Axis (cm) | Short Axis (cm) | Volume (cm$^3$) | Adjust. to 75% | Weight (kg) | Estimated Wave ht. (m) |
|---|---|---|---|---|---|---|---|---|
| 1 | 0 | 100 | 60 | 35 | 210,000 | 157,500 | 402 | 8.3 |
| 2 | +100 | 76 | 46 | 36 | 125,856 | 94,374 | 241 | 6.3 |
| 3 | +250 | 85 | 63 | 23 | 123,165 | 92,374 | 236 | 7.1 |
| 4 | +200 | 82 | 65 | 34 | 181,220 | 135,914 | 347 | 6.8 |
| 5 | +240 | 100 | 50 | 26 | 130,000 | 97,500 | 249 | 8.3 |
| 6 | +140 | 82 | 45 | 36 | 132,840 | 99,630 | 254 | 6.8 |
| 7 | +150 | 42 | 37 | 36 | 55,944 | 41,958 | 107 | 3.5 |
| 8 | +210 | 66 | 33 | 25 | 54,450 | 40,838 | 104 | 5.5 |
| 9 | +70 | 57 | 30 | 23 | 39,330 | 29,497 | 75 | 4.7 |
| 10 | +230 | 67 | 38 | 21 | 53,466 | 40,100 | 102 | 5.6 |
| 11 | +70 | 70 | 34 | 25 | 59,500 | 44,625 | 114 | 5.8 |
| 12 | +50 | 69 | 43 | 17 | 50,439 | 37,829 | 96 | 5.7 |
| 13 | +100 | 64 | 36 | 24 | 55,296 | 41,472 | 106 | 5.3 |
| 14 | +100 | 69 | 35 | 32 | 77,280 | 57,960 | 148 | 5.7 |
| 15 | +130 | 81 | 38 | 23 | 70,794 | 53,096 | 135 | 6.7 |
| 16 | +220 | 84 | 61 | 48 | 245,952 | 184,464 | 470 | 7.0 |
| 17 | +370 | 52 | 51 | 22 | 58,344 | 43,758 | 112 | 4.3 |
| 18 | +260 | 78 | 75 | 23 | 134,550 | 100,913 | 257 | 6.5 |
| 19 | +170 | 89 | 55 | 49 | 239,855 | 179,891 | 459 | 7.4 |
| 20 | +230 | 131 | 43 | 41 | 230,953 | 173,215 | 442 | 10.9 |
| 21 | +100 | 63 | 35 | 25 | 55,125 | 41,344 | 105 | 5.2 |
| 22 | +100 | 5 | 39 | 19 | 41,49 | 31,122 | 79 | 4.7 |
| 23 | +420 | 58 | 35 | 25 | 50,750 | 38,063 | 97 | 4.8 |
| 24 | +100 | 75 | 31 | 28 | 65,10 | 48,825 | 125 | 6.2 |
| 25 | +250 | 50 | 27 | 26 | 35,100 | 26,325 | 67 | 4.2 |
| Average | +170 | 74 | 44 | 29 | 103,072 | 77,303 | 197 | 6.3 |

**Table 2.** Quantification of boulder size, volume and estimated weight from coastal bar samples through Transect 1b (continuation from 1a east of El China). The laboratory result for density at 2.55 gm/cm$^3$ is applied uniformly to all samples in order to calculate wave height for each boulder.

| Sample | Distance to Next (cm) | Long Axis (cm) | Intermediate Axis (cm) | Short Axis (cm) | Volume (cm$^3$) | Adjust. to 75% | Weight (kg) | Estimated Wave ht. (m) |
|---|---|---|---|---|---|---|---|---|
| 1 | 0 | 73 | 31 | 23 | 52,049 | 39,037 | 100 | 6.1 |
| 2 | 300 | 47 | 29 | 25 | 34,075 | 25,556 | 65 | 3.9 |
| 3 | 130 | 59 | 25 | 18 | 26,550 | 19,913 | 51 | 4.9 |
| 4 | 170 | 86 | 32 | 18 | 49,536 | 66,048 | 168 | 7.3 |
| 5 | 300 | 49 | 39 | 26 | 49,686 | 66,248 | 169 | 4.1 |
| 6 | 30 | 44 | 42 | 39 | 72,072 | 96,096 | 245 | 3.7 |
| 7 | 300 | 36 | 25 | 23 | 20,700 | 15,525 | 40 | 3 |
| 8 | 210 | 42 | 28 | 23 | 27,048 | 20,286 | 52 | 3.5 |
| 9 | 160 | 72 | 39 | 29 | 81,432 | 61,074 | 156 | 6 |
| 10 | 160 | 48 | 25 | 25 | 30,000 | 22,500 | 57 | 4 |
| 11 | 100 | 95 | 40 | 29 | 110,200 | 82,650 | 211 | 7.9 |
| 12 | 190 | 60 | 36 | 19 | 41,040 | 30,780 | 78 | 5 |
| 13 | 110 | 59 | 32 | 22 | 41,536 | 31,152 | 79 | 4.9 |
| 14 | 170 | 62 | 45 | 28 | 78,120 | 58,590 | 149 | 5.2 |
| 15 | 330 | 59 | 41 | 29 | 70,151 | 52,613 | 134 | 4.9 |
| 16 | 200 | 47 | 23 | 17 | 18,377 | 13,783 | 35 | 3.9 |
| 17 | 200 | 62 | 36 | 33 | 73,656 | 55,242 | 141 | 5.2 |
| 18 | 170 | 78 | 29 | 25 | 56,550 | 42,412 | 108 | 6.5 |
| 19 | 230 | 63 | 40 | 25 | 63,000 | 47,250 | 120 | 5.2 |
| 20 | 170 | 70 | 49 | 37 | 126,910 | 95,183 | 243 | 5.8 |
| 21 | 100 | 74 | 43 | 41 | 130,462 | 97,842 | 250 | 6.3 |
| 21 | 220 | 52 | 34 | 18 | 31,824 | 23,868 | 61 | 4.3 |
| 23 | 180 | 46 | 33 | 23 | 34,914 | 26,186 | 67 | 3.8 |
| 24 | 100 | 75 | 28 | 22 | 46,200 | 34,650 | 88 | 6.2 |
| 25 | 150 | 53 | 48 | 27 | 68,688 | 51,516 | 131 | 4.4 |
| Average | 175 | 60 | 35 | 26 | 57,391 | 47,040 | 120 | 5 |

**Table 3.** Quantification of boulder size, volume and estimated weight from coastal bar samples through Transect 1c (continuation from 1b east of El China). The laboratory result for density at 2.55 gm/cm$^3$ is applied uniformly in order to calculate wave height for each boulder.

| Sample | Distance to Next (cm) | Long Axis (cm) | Intermediate Axis (cm) | Short Axis (cm) | Volume (cm$^3$) | Adjust. to 75% | Weight (kg) | Estimated Wave ht. (m) |
|---|---|---|---|---|---|---|---|---|
| 1 | 0 | 45 | 30 | 17 | 22,950 | 17,213 | 44 | 3.7 |
| 2 | 250 | 40 | 21 | 15 | 12,600 | 9,450 | 24 | 3.3 |
| 3 | 120 | 58 | 22 | 19 | 24,244 | 18,183 | 46 | 4.8 |
| 4 | 130 | 60 | 40 | 18 | 43,200 | 32,400 | 83 | 5 |
| 5 | 170 | 86 | 35 | 33 | 99,330 | 74,498 | 190 | 7.3 |
| 6 | 200 | 61 | 28 | 24 | 40,992 | 30,744 | 78 | 5.1 |
| 7 | 110 | 65 | 35 | 16 | 36,400 | 27,300 | 70 | 5.4 |
| 8 | 180 | 55 | 25 | 24 | 33,000 | 24,750 | 63 | 4.6 |
| 9 | 260 | 69 | 42 | 20 | 57,960 | 43,470 | 111 | 5.7 |
| 10 | 250 | 49 | 37 | 18 | 32,634 | 24,476 | 62 | 4.1 |
| 11 | 250 | 44 | 34 | 33 | 49,368 | 37,026 | 94 | 3.7 |
| 12 | 330 | 52 | 36 | 29 | 54,288 | 40,716 | 104 | 4.3 |
| 13 | 180 | 59 | 18 | 18 | 19,116 | 31,152 | 37 | 4.9 |
| 14 | 320 | 36 | 12 | 17 | 7,344 | 5,508 | 14 | 3 |
| 15 | 80 | 36 | 28 | 17 | 17,136 | 12,852 | 33 | 3 |
| 16 | 240 | 39 | 22 | 12 | 10,296 | 7,722 | 20 | 3.2 |
| 17 | 230 | 38 | 20 | 18 | 13,680 | 10,260 | 26 | 3.2 |
| 18 | 270 | 49 | 31 | 13 | 19,747 | 42,412 | 108 | 4.1 |
| 19 | 130 | 50 | 36 | 18 | 63,000 | 14,810 | 38 | 4.2 |
| 20 | 230 | 80 | 50 | 29 | 32,400 | 24,300 | 62 | 6.7 |
| 21 | 110 | 49 | 25 | 25 | 116,000 | 87,000 | 222 | 4.1 |
| 22 | 160 | 37 | 22 | 13 | 30,625 | 22,969 | 59 | 3.1 |
| 23 | 300 | 43 | 28 | 18 | 21,672 | 16,254 | 41 | 3.6 |
| 24 | 250 | 48 | 24 | 15 | 17,280 | 12,960 | 33 | 4 |
| 25 | 60 | 79 | 37 | 32 | 93,536 | 70,152 | 179 | 6.6 |
| Average | 198 | 53 | 30 | 20 | 38,752 | 29,064 | 74 | 44 |

**Table 4.** Quantification of boulder size, volume and estimated weight from coastal bar samples through Transect 1d (continuation from 1c east of El China). The laboratory result for density at 2.55 gm/cm$^3$ is applied uniformly in order to calculate wave height for each boulder.

| Sample | Distance to Next (cm) | Long Axis (cm) | Intermediate Axis (cm) | Short Axis (cm) | Volume (cm$^3$) | Adjust. to 75% | Weight (kg) | Estimated Wave ht. (m) |
|---|---|---|---|---|---|---|---|---|
| 1 | 0 | 45 | 31 | 25 | 34,875 | 26,156 | 67 | 3.7 |
| 2 | 320 | 51 | 30 | 24 | 36,720 | 27,540 | 70 | 4.2 |
| 3 | 330 | 49 | 26 | 23 | 29,320 | 21,990 | 56 | 4.1 |
| 4 | 100 | 38 | 19 | 17 | 12,274 | 9,206 | 23 | 3.2 |
| 5 | 180 | 40 | 21 | 18 | 15,120 | 11,340 | 29 | 3.3 |
| 6 | 210 | 43 | 16 | 14 | 9,632 | 7,224 | 18 | 3.6 |
| 7 | 190 | 47 | 19 | 17 | 15,181 | 11,386 | 29 | 1.9 |
| 8 | 200 | 59 | 37 | 26 | 56,758 | 42,569 | 109 | 4.9 |
| 9 | 220 | 57 | 20 | 20 | 22,800 | 17,100 | 44 | 4.7 |
| 10 | 110 | 57 | 27 | 16 | 24,624 | 18,468 | 47 | 4.7 |
| 11 | 170 | 47 | 23 | 14 | 15,134 | 11,351 | 29 | 3.9 |
| 12 | 30 | 38 | 20 | 19 | 14,440 | 10,830 | 28 | 3.2 |
| 13 | 210 | 96 | 56 | 20 | 107,520 | 80,640 | 206 | 8 |
| 14 | 180 | 76 | 46 | 32 | 111,872 | 83,904 | 214 | 6.3 |
| 15 | 120 | 54 | 42 | 30 | 68,040 | 51,030 | 130 | 4.5 |
| 16 | 140 | 52 | 25 | 24 | 31,200 | 23,400 | 60 | 4.3 |
| 17 | 40 | 72 | 35 | 18 | 45,360 | 34,020 | 87 | 6 |
| 18 | 200 | 57 | 26 | 20 | 29,640 | 22,230 | 57 | 4.7 |
| 19 | 100 | 67 | 41 | 30 | 82,410 | 61,808 | 158 | 5.6 |
| 20 | 250 | 59 | 30 | 21 | 37,170 | 37,170 | 71 | 4.9 |
| 21 | 380 | 76 | 40 | 26 | 79,040 | 59,280 | 151 | 6.3 |
| 22 | 20 | 74 | 40 | 38 | 112,480 | 84,360 | 215 | 6.2 |
| 23 | 120 | 51 | 39 | 27 | 53,703 | 40,277 | 103 | 4.2 |
| 24 | 170 | 55 | 42 | 30 | 69,300 | 51,975 | 133 | 4.6 |
| 25 | 270 | 78 | 50 | 31 | 120,900 | 90,675 | 231 | 6.5 |
| Average | 170 | 58 | 32 | 23 | 49,421 | 37,437 | 95 | 4.8 |

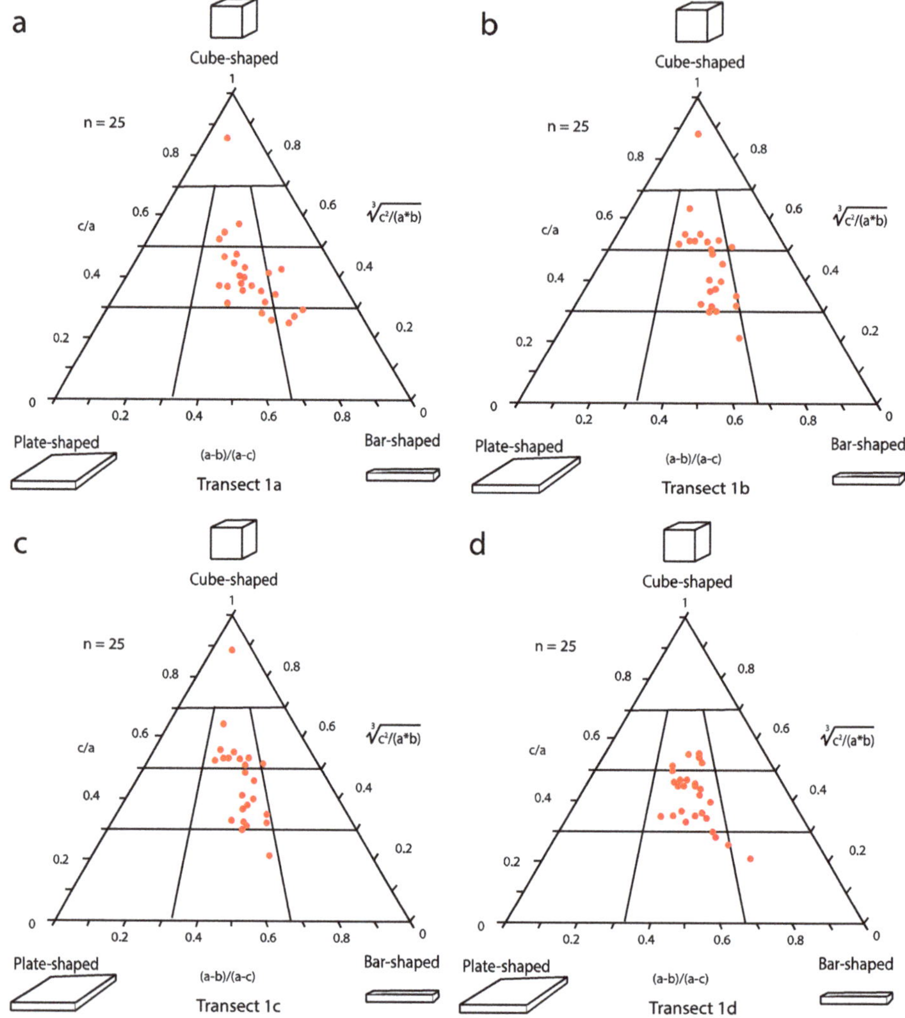

**Figure 7.** Set of four triangular Sneed-Folk diagrams used to appraise variations in boulder shape on barrier #1; (**a**) Trend for boulders from Transect 1a; (**b**) Trend for boulders from Transect 1b; (**c**) Trend for boulders from Transect 1c; (**d**) Trend for boulders from Transect 1d. Note the similarity in slopes from sample to sample.

Comparable data from barrier #2 collected along two consecutive transects of 50 m each are registered in Tables 5 and 6. Individual boulders from the upper intertidal zone of barrier #2 are shown by data points plotted in a pair of Sneed-Folk triangular diagrams. The trends expressed in Figure 8a,b are similar both to one another, as well as to those found in Figure 7a–d. That is, the overprint of a common pattern immerges in which the constituent boulders entrained in both barriers trend toward shapes that are more elongated and not at all plate-shaped.

**Table 5.** Quantification of boulder size, volume and estimated weight from coastal bar samples collected from transect 2a. (north of Cerro Enfermería). The laboratory result for density at 2.55 gm/cm$^3$ is applied uniformly in order to calculate wave height for each boulder.

| Sample | Distance to Next (cm) | Long Axis (cm) | Intermediate Axis (cm) | Short Axis (cm) | Volume (cm$^3$) | Adjust. to 75% | Weight (kg) | Estimated Wave ht. (m) |
|---|---|---|---|---|---|---|---|---|
| 1 | 0 | 62 | 28 | 26 | 45,136 | 33,852 | 86 | 5.2 |
| 2 | 220 | 73 | 34 | 21 | 52,122 | 39,092 | 100 | 6.1 |
| 3 | 210 | 118 | 63 | 42 | 312,228 | 234,171 | 597 | 9.8 |
| 4 | 170 | 86 | 58 | 33 | 164,604 | 123,453 | 315 | 7.2 |
| 5 | 170 | 48 | 32 | 28 | 43,008 | 32,256 | 82 | 4 |
| 6 | 160 | 48 | 38 | 29 | 52,896 | 39,672 | 101 | 4 |
| 7 | 10 | 58 | 43 | 27 | 67,338 | 50,504 | 129 | 4.8 |
| 8 | 210 | 61 | 48 | 45 | 131,760 | 98,820 | 252 | 5.1 |
| 9 | 380 | 90 | 53 | 23 | 109,710 | 82,283 | 210 | 7.5 |
| 10 | 220 | 75 | 58 | 29 | 126,150 | 94,613 | 241 | 6.2 |
| 11 | 320 | 55 | 38 | 35 | 73,150 | 54,863 | 140 | 4.6 |
| 12 | 180 | 82 | 51 | 22 | 92,004 | 69,003 | 176 | 6.8 |
| 13 | 220 | 69 | 25 | 17 | 29,325 | 21,994 | 56 | 5.7 |
| 14 | 80 | 78 | 52 | 25 | 101,400 | 76,050 | 194 | 6.5 |
| 15 | 100 | 78 | 34 | 25 | 66,300 | 49,725 | 127 | 6.5 |
| 16 | 80 | 64 | 43 | 20 | 55,040 | 41,280 | 105 | 5.3 |
| 17 | 120 | 59 | 41 | 23 | 55,637 | 41,728 | 106 | 4.9 |
| 18 | 80 | 80 | 34 | 25 | 68,000 | 51,000 | 130 | 6.7 |
| 19 | 110 | 65 | 39 | 33 | 83,655 | 62,741 | 160 | 5.4 |
| 20 | 230 | 65 | 45 | 17 | 49,725 | 37,294 | 95 | 5.4 |
| 21 | 180 | 88 | 39 | 35 | 120,120 | 90,090 | 230 | 7.3 |
| 22 | 150 | 72 | 44 | 22 | 69,696 | 52,072 | 133 | 6 |
| 23 | 140 | 73 | 60 | 22 | 96,360 | 72,270 | 184 | 6.1 |
| 24 | 180 | 67 | 31 | 20 | 41,540 | 31,155 | 79 | 5.6 |
| 25 | 230 | 64 | 39 | 25 | 62,400 | 46,800 | 119 | 5.3 |
| Average | 166 | 71 | 43 | 27 | 86,772 | 65,071 | 166 | 5.9 |

**Table 6.** Quantification of boulder size, volume and estimated weight from coastal bar samples from transect 2b. (north of Cerro Enfernmera). The laboratory result for density at 2.55 gm/cm$^3$ is applied uniformly in order to calculate wave height for each boulder.

| Sample | Distance to Next (cm) | Long Axis (cm) | Intermediate Axis (cm) | Short Axis (cm) | Volume (cm$^3$) | Adjust. to 75% | Weight (kg) | Estimated Wave ht. (m) |
|---|---|---|---|---|---|---|---|---|
| 1 | 0 | 118 | 52 | 48 | 294,528 | 220,896 | 563 | 9.8 |
| 2 | 300 | 56 | 37 | 23 | 47,656 | 35,742 | 91 | 4.7 |
| 3 | 110 | 51 | 24 | 17 | 20,808 | 15,606 | 40 | 4.2 |
| 4 | 260 | 79 | 49 | 22 | 85,162 | 63,872 | 163 | 6.6 |
| 5 | 240 | 50 | 40 | 22 | 44,000 | 33,000 | 84 | 4.2 |
| 6 | 110 | 51 | 35 | 26 | 46,410 | 34,808 | 89 | 4.2 |
| 7 | 60 | 66 | 40 | 17 | 44,880 | 33,660 | 86 | 5.5 |
| 8 | 100 | 67 | 38 | 30 | 76,380 | 57,285 | 146 | 5.6 |
| 9 | 150 | 66 | 43 | 26 | 73,788 | 55,341 | 141 | 5.5 |
| 10 | 170 | 58 | 44 | 35 | 89,320 | 66,990 | 171 | 4.8 |
| 11 | 70 | 86 | 49 | 31 | 130,634 | 97,976 | 250 | 7.2 |
| 12 | 250 | 63 | 44 | 20 | 55,440 | 41,580 | 106 | 5.2 |
| 13 | 60 | 75 | 45 | 18 | 60,750 | 45,563 | 116 | 6.1 |
| 14 | 120 | 63 | 31 | 27 | 52,731 | 39,548 | 101 | 5.3 |
| 15 | 280 | 86 | 45 | 21 | 81,270 | 60,953 | 155 | 5.3 |
| 16 | 90 | 60 | 37 | 35 | 116,550 | 87,413 | 223 | 7.2 |
| 17 | 320 | 56 | 35 | 31 | 60,760 | 45,570 | 116 | 5 |
| 18 | 0 | 46 | 34 | 22 | 34,408 | 25,806 | 66 | 4.7 |
| 19 | 300 | 61 | 34 | 27 | 55,998 | 41,999 | 107 | 3.8 |
| 20 | 100 | 70 | 44 | 28 | 86,240 | 64,680 | 165 | 5.1 |
| 21 | 470 | 89 | 58 | 42 | 216,804 | 162,603 | 415 | 5.8 |
| 22 | 180 | 59 | 33 | 26 | 50,622 | 37,967 | 97 | 7.4 |
| 23 | 120 | 59 | 31 | 17 | 31,093 | 23,320 | 59 | 4.9 |
| 24 | 380 | 48 | 31 | 24 | 35,712 | 26,784 | 68 | 4 |
| 25 | 360 | 70 | 42 | 21 | 61,740 | 46,305 | 118 | 5.8 |
| Average | 184 | 66 | 40 | 26 | 78,147 | 58,611 | 149 | 5.5 |

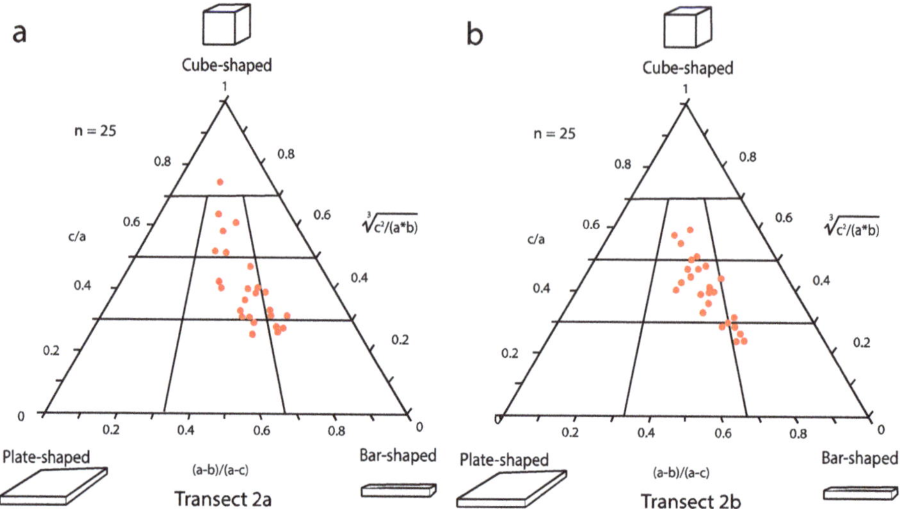



**Figure 8.** Pair of triangular Sneed-Folk diagrams used to appraise variations in boulder shape on barrier #2; (**a**) Trend for boulders from Transect 2a; (**b**) Trend for boulders from Transect 2b. Note similarities in slopes with those from barrier #1 in Figure 7.

### 4.5. Analysis of Boulder Sizes

Variations in boulder size as a function of maximum and intermediate length drawn from the data sets for barrier #1 (Tables 1–4) are plotted separately for each of four transects using bar graphs. In this case, the stacked succession of graphs in Figure 9a–d show that the extreme outlier in maximum boulder size occurs in the first transect nearest the sea cliffs on Cerro El Chino. Boulders in the size class between 41 and 55 cm in diameter become more abundant and those in the size class between 56 and 70 cm in diameter are fewer in number. Otherwise, the size distributions remain fairly consistent from transect 1b through transect 1d. However, a deviation signaling a minor reversal in the size class between 71 and 85 cm appears in a comparison of Figure 9c,d. Such a reversal could imply a change in the direction of boulder source coming from the intermediate islet to the south. The numbers involved are small.

Boulder sizes along the intermediate axis from transects 1a through 1d are plotted in the stacked bar graphs from Figure 9e–h. Not surprisingly, the outlier in extreme length is found in Figure 8e representing the transect nearest the sea cliffs on Cerro El Chino. Otherwise, boulders in the size class between 26 and 40 cm are fairly consistent in number through the four transects. Data comparing the relative frequency of boulders in different size classes from barrier #2 are plotted as bar graphs in Figure 10a,b for the long axis and Figure 10c,d for the intermediate axis. Differences between the two samples are few. Most noticeable is the diminishment of boulders in the size class between 71 and 85 cm from transect 2a to 2b in regard to length across the long axis. Otherwise, apparent differences in size variation tend to be minimal. Again, a minor deviation appears in close comparison among Figure 9f–h, with particular reference to the size class between 41 and 55 cm. Moreover, a single large boulder in the size class between 56 and 70 cm is re-established.

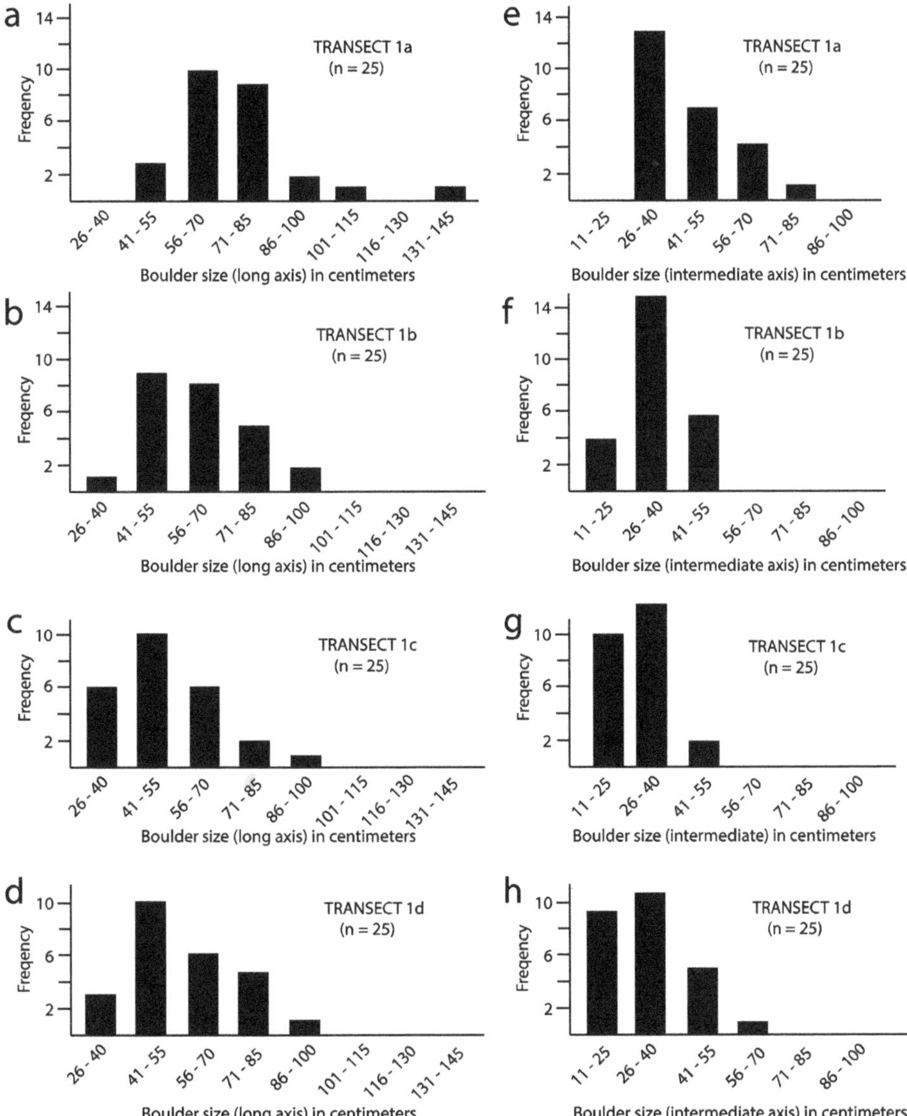

**Figure 9.** Parallel sets of bar graphs used to appraise variations in the long and intermediate axes on boulders from barrier #1; (**a**) Long axis from boulders in Transect 1a; (**b**) Long axis from boulders in Transect 1b; (**c**) Long axis from boulders in Transect 1c; (**d**) Long axis from boulders in Transect 1d; (**e**) intermediate axes from boulders in Transect 1a; (**f**); Intermediate axis from boulders in Transect 1b; (**g**); Intermediate axis from boulders in Transect 1c; (**h**) Intermediate axis from boulders in Transect 1d.

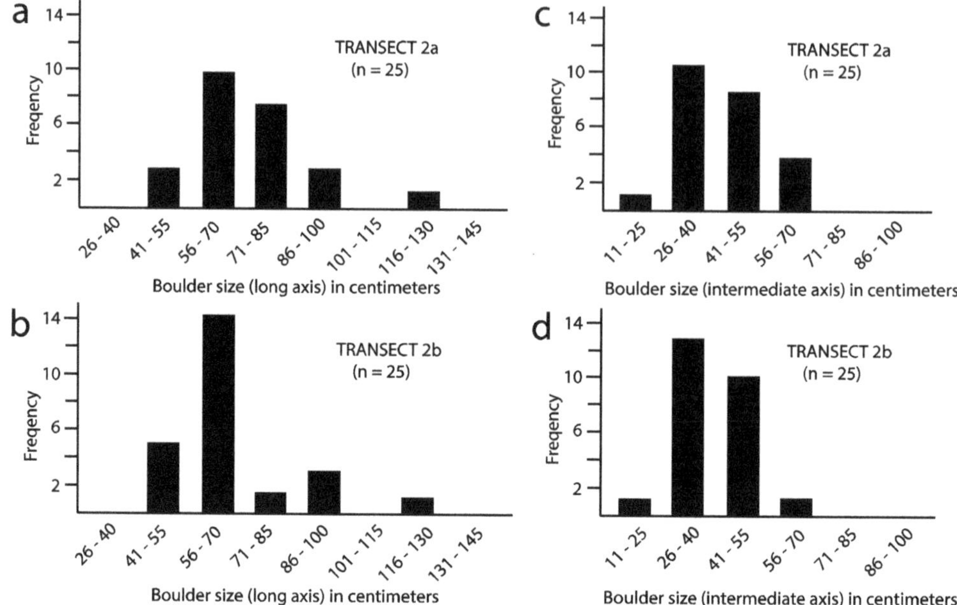

**Figure 10.** Parallel sets of bar graphs used to appraise variations in the log and intermediate axes on boulders from barrier #2; (**a**) Long axis from boulders in Transect 2a; (**b**) Long axis from boulders in Transect 2b; (**c**) Intermediate axis from boulders in Transect 2a; (**d**) Intermediate axis from boulders in Transect 2b.

### 4.6. Estimation of Wave Heights

A summary of key data is provided (Table 7), pertaining to average boulder size and maximum boulder size from the four transects in barrier #1 and two transects in barrier #2 as correlated with weight calculated on the basis of specific gravity for andesite. These data are applied to estimate the wave heights required to transport boulders from the bedrock source in sea cliffs to their resting place embedded in the natural breakwaters. The estimated wave height needed to move the largest bolder encountered in Transect 1a amounts to 7 m, although the average computed for the 25 boulders in that sample amounts to 6.3 m. Average boulder size generally declines through the four transects from barrier #1, as does the average wave height estimated to move those boulders. Regarding barrier #2, the average boulder size declines from transect 2a to 2b and so does the average wave height estimated to move those boulders. However, there is no difference between the largest boulders from the two transects with respect to maximum size and estimated wave height required to shift those boulders.

**Table 7.** Summary data from Tables 1–6 showing maximum boulder size and estimated weight compared to the average values for all boulders (N = 25) from each of transects 1–6 together with calculated values for wave heights estimated as necessary for boulder mobility.

| Tran-Sect | Number of Samples | Average Boulder Size (cm$^3$) | Average Bolder Weight (kg) | Estimated Average Wave ht. (m) | Max. Boulder Size (cm$^3$) | Max. Bolder Weight (kg) | Estimated Wave Height (m) |
|---|---|---|---|---|---|---|---|
| 1a | 25 | 77,303 | 197 | 6.3 | 179,891 | 470 | 7 |
| 1b | 25 | 47,040 | 120 | 5 | 97,842 | 250 | 6.3 |
| 1c | 25 | 57,391 | 74 | 4.4 | 87,000 | 222 | 4.1 |
| 1d | 25 | 37,437 | 95 | 4.8 | 90,675 | 215 | 6.2 |
| 2a | 25 | 65,071 | 166 | 5.9 | 234,171 | 597 | 9.8 |
| 2b | 25 | 58,611 | 149 | 5.5 | 220,896 | 563 | 9.8 |

*4.7. Implications of Geomorphologic Modeling*

The extent of topographic asymmetry across the outer bulwark of lands fringing Puerto Escondido (Figure 4) invites geomorphologic modeling aimed at accounting for the amount of rock volume lost due to coastal recession. This exercise targets the un-named islet between barriers #1 and #2, where the object's shape is relatively small and simple. Inherent in the model is the assumption that a body of bedrock with uniform composition starts out having more balanced proportions at the commencement of physical erosion. Three stages are depicted graphically in the model (Figure 11). First, the islet's present-day topography is laid out on a regular grid and a diagonal line is drawn such that the asymmetry is segregated to one side (Figure 11a). The area within each successive line of topography is estimated separately and thereafter, volume may be calculated through addition in discrete topographic intervals much like adding layers in a tiered wedding cake. Following this procedure, the bulk volume of the andesite islet is found to be roughly 13.85 million cubic meters. In stage 2 (Figure 11b), those topographic lines with the closest spacing are erased. In stage 3 (Figure 11c), the size of a former islet is reconstructed by redrawing topographic lines as more evenly spaced. Thereafter, the same procedure may be followed to arrive at the bulk volume of the enlarged islet. In this way, the former islet is found to have started with a bulk volume of 19.25 million cubic meters. Subtracting present-day volume from the reconstituted volume, the original islet is argued to have lost 5.4 million cubic meters. That amount is crudely equivalent to 25% of the islet's former volume due to an imbalance of coastal erosion on its exposed seaward flank.

It is essential to point out that the dividing line applied in the model (Figure 11a–c) is not a fault line. However, a hidden fault line now underwater is projected over a distance of 4 km along the outer coast but also notably parallel to the inland valley fault on the west side of Puerto Escondido (Figure 4). Separate calculations on the volume of unconsolidated materials in the two barriers is based on the length of each structure, its average width, and depth. The last is problematic to appraise but, using an assumed value of 5.5 m, can be no greater than the maximum depth of the lagoon behind. Hence, the contents entrained in the longer barrier #1 may amount to 42.5 million cubic meters. The shorter barrier #2 holds no less than 23 million cubic meters of transported pebbles, cobbles, and boulders. In effect, the barriers that insure the sheltered inner harbor at Puerto Escondido might be removed and restored many times over based on the volume of solid bedrock deleted by coastal erosion along an original fault scarp now significantly recessed.

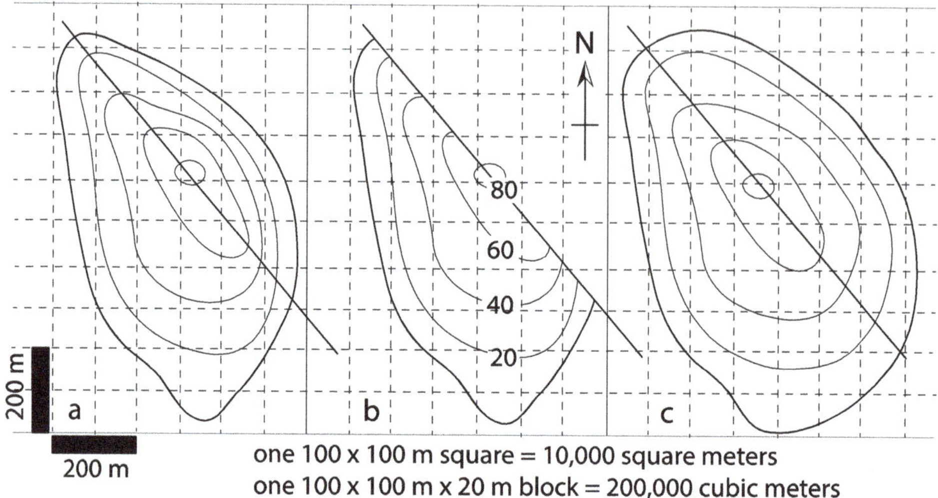


**Figure 11.** Progression of stages in geomorphological modeling with respect to the un-named islet between barriers #1 and #2; (**a**) Topography of present day islet with dividing line showing disparity between the gentler lagoon side and exposed cliff face on the open sea; (**b**) Erasure of the outer cliff face; (**c**) Restoration of past topographic gradient on the seaward face showing a better match with the sheltered side.

## 5. Discussion

### 5.1. Time Constraints on Barrier Origin

Relationships drawn from paleontological and geological evidence place time limitations on the origin of the two barriers crucial to the maintenance of shelter at Puerto Escondido. Upper Pleistocene deposits denoted by a shell drape over a sizable area in the hills west of the inner harbor (Figure 4) confirm that normal seawater circulated to that spot approximately 125,000 years ago during the last interglacial epoch. Fossil mollusks including species such as the dominant *Chione californiensis* and less abundant *Turbo fluctuosus* that still live in the Gulf of California today [18], are reliably taken as evidence for intertidal conditions at that locality. Sea level stood approximately 6 m higher compared to now [16,17], but the present elevation of the shell drape also reflects additional tectonic uplift. Higher global sea level probably enhanced marine circulation at this spot. Moreover, emplacement of the Upper Pleistocene deposit must have occurred prior to development of barriers #1 and #2. The fault traced through the narrow valley from Bahía Juncalito indicates that local uplift boosted the shell drape to a higher elevation than typically found in coeval shell drapes on 12-m terraces many places elsewhere along peninsular gulf shores [15]. In fact, the nearest expression of marine terraces cut in andesite bedrock occurs within sight of Puerto Escondido only 7 km across the Carmen Passage at the southern end of Isla del Carmen (Figure 12).

No traces of marine terraces are found along the outer shores at Puerto Escondido facing Isla del Carmen. Given the amount of coastal retreat implicated by the geomorphological model, such terraces may have existed but were entirely erased by coastal erosion. If true, the earliest development of barriers #1 and #2 occurred after the end of the Pleistocene in Holocene time. However, the physical juxtaposition on opposite sides of the Carmen Passage raises the question why marine terraces survived on neighboring Isla del Carmen but not the outer coast at Puerto Escondido? The answer likely lies in the principle sources of coastal erosion still taking place in the Gulf of California, today.

**Figure 12.** View east from Puerto Escondido across the Carmen Passage showing distinct marine terraces cut in the southwest side of Isla del Carmen. Motor boat and wake for scale at left, center.

## 5.2. Energy Sources Affecting Barrier Development

As discussed in our earlier work [1,2], the potential range of dynamic influences capable of shore erosion in the Gulf of California includes tidal action, long-shore currents related to strong seasonal winds, alleged tsunamis, and hurricanes. Tidal influence is especially strong in the far northern part of the gulf, where maximum amplitudes of 12 m are recorded. The tidal range around the central gulf is far less, approximately 2.75 m [19]. Tides of this magnitude transport coarse sand, but have little or no effect on rocky shores. South-directed sea swells with an amplitude of 2 m and wavelength of 10 m are not unusual during episodes of strong winds in the Carmen Passage that play out episodically between November and May [6,7]. Such prevailing winter winds stimulate long-shore currents that flow parallel to the gulf shores, or otherwise result in wave refraction around obstructing islands or headlands [20]. The energy generated by such currents is capable of moving pebbles and smaller cobbles.

Sea storms of lesser intensity are expected to shift sand, pebbles, and even cobbles entrained in a natural barrier. In part, the overall decrease in boulder size from north to south from transect 1a to 1c and from transect 2a to 2b is related to littoral drift especially during the winter season when a strong north-south wind is common. It can be argued, however, that only those episodic storms of hurricane intensity generate sufficient energy to shift large boulders close to a metric ton in weight. During the Pacific Ocean hurricane season between the months of May and November, between 25 and 30 tropical depression originate off the southwest coast of mainland Mexico [8], but few diverge from an outward path to enter the Gulf of California. The incidence of hurricane activity in the gulf region increases every 6 to 8 years during El Niño events. Hurricane Odile in September 2014, for example, was filmed in action as it pounded the rocky coast near the Almeja CBD north of Loreto with waves that impacted sea cliffs at a height 8 m above normal [2]. Clocking wind speeds of 113 km/hr by the time it reached that far into the Gulf of California, wind bands rotating counter clockwise were strong enough to generate wave surge that lashed the coast initially from east to west. As the storm migrated northward, wind direction and wave surge shifted more to a direction from northeast to southwest. Such a pattern fits the predicted scenario of rocky-shore erosion and transfer of large boulders to the barrier seawalls at Puerto Escondido. In particular, oblong blocks of andesite already fallen from the unstable sea cliffs at Cerro El Chino and the un-named islet between barriers #1 and #2 would be pushed southward and eventually entrained in those barriers (Figures 9 and 10). Marine terraces on SW Carmen (Figure 12) would be sheltered from west-moving storm bands, and therefore avoid excessive erosion.

In theory, a large tsunami with a run-up of several meters would be energetic enough to breach the barriers protecting Puerto Escondido. In addition, the return outwash of coastal sediments dislodged during a tsunami should be transported seaward. However, the probability that a tsunami struck anywhere within the Gulf of California during the Holocene is nil, even though a recent interpretation of sedimentary deposits by McCloskey et al. (2015) for the lower Gulf of California [21] suggested a geological framework and seismic mechanisms for its interpretation. In the Alfonso Basin off La Paz, Gorsline et al. (2000) described only minor discharges from tributary coastal canyons that carried a high proportion of coarse-grained sand trapped on marine shelves, but blocked from supplying turbidites at a volume of a basin-wide magnitude [22]. Hence, those sediments reaching the basin floor probably were produced by seismically generated slope failures of silty clay deposits. The distribution of the dated turbidites and a slip face in a box core from the landward slope, indicate a source on the landward depositional slope of the fault-bounded basin. Comparable discontinuities of the same age also are reported from the east side of the gulf in the Guaymas area farther north [22].

Most earthquakes in the lower Gulf of California are generated by transform faults [23]. Fletcher and Mungia (2000) indicate that such a level of seismicity falls along different strands in a major system of normal faults extending at least 300 km along strike to define the western limit of the Gulf Extensional Province [24]. The dominant normal faults controlled distribution of Neogene basins active during middle to late Miocene times. Structural analysis of secondary faults in the southern gulf segment reveals that fault populations are Pliocene to Holocene in age and represented by mixed normal and dextral-normal faults with a bulk extension direction of west-northwest–east-southeast [25]. From the perspective of regional tectonics summarized above, it is evident that a major earthquake is not responsible for the presence of turbidites within the Alfonso basin at least during recent times.

Based on the experience of the junior author (J.L.-V.), the extraneous evidence shown by McCloskey et al. (2015, their Figure 5c) relates to a kitchen midden and not a tsunami deposit. Tsunami events are well documented outside the Gulf of California far to the south on the Mexican mainland at Jalisco [26], but such events result from deep-seated earthquakes (magnitude 7.7 or greater) associated with an active subduction zone where the Rivera lithospheric plate meets the continental mainland.

*5.3. Comparisons with Other Coastal Boulder Deposits*

Our previous contributions on Holocene boulder accumulations within Mexico's Gulf of California conform to the normal definition of coastal boulder deposits, where CBDs occur either at the top of sea cliffs or next to sea cliffs with well-developed bedrock stratification and jointing that corresponds to the dimensions of rocks loosened by wave impact. Metric-ton blocks of Pliocene limestone sit atop 12 m high cliffs on Isla del Carmen (Figure 1b, locality 2) from which they were peeled away [2]. Similar-size rhyolite boulders at Ensenada Almeja (Figure 1b, locality 3) occur at sea level adjacent to the sea cliffs from which they were extracted [3]. Elsewhere, major CBDs occur atop sea cliffs in northern France [27] and western Ireland [28,29]. Mega-boulders left high above sea level but close to the parent bed rock from which they were eroded also are known from the Bahamas and Bermuda [30,31]. It is debated to what extent super-waves and hurricanes are responsible for these CBDs, fueling ongoing controversy over the growing threat of global warming. In the case of giant blocks derived from low limestone cliffs on Calicoan Island in the Philippines [32], the run-up of waves exceeding 15 m inland is linked directly to the impact of Super Typhoon Haiyan in November 2013. A related but somewhat different phenomenon concerns the detachment of large boulders from the seafloor and onshore transferal to low rocky shores. The recent study by Biolchi et al. (2019) fits this category in relation to movement of limestone boulders in the northern Adriatic Sea onto the Premantura (Kamenjak) Promontory in Croatia [33]. In this case, however, the coastline is formed by a low-angle rocky shore that is more like a ramp in configuration than a sea cliff.

The barriers that close off the inner harbor at Puerto Escondido are natural breakwaters formed by a mixture of pebbles, cobbles, and boulders that trace back to single sources of bed rock in exposed sea cliffs. Those sea cliffs are steeply inclined (50° to 55°) and are unstable. Andesite layers within the bedrock are tilted at a high angle dipping westward and rock falls leave fresh material at the base of the cliffs. Essentially, a combination of long-shore currents and storm waves harvest the materials and carry them southward where they are entrained in the linear barriers. A more appropriate term for this kind of feature is a barrier boulder deposit (BBD). This term is a good fit with the many bars formed by andesite cobbles and boulders that close off lagoons on Isla Angel de la Guarda in the upper Gulf of California—some of which extend for as much as 1.25 km [34]. The aerial photo from Johnson et al. (2019, their Figure 10) illustrates two such barriers and others in the process of extension from the nearest bedrock source [2]. Granite is another major rock type from which rocky shores are formed in the Gulf of California. Eroded granite boulders at Bahía San Antonio (Figure 1, locality 4) are encrusted by Upper Pleistocene fossils representing an intertidal biota preserved in growth position [35]. Close to granite bedrock at Punta San Antonio, the scenario corresponds well with the BBD concept. The potential for development of a BBD appears to be less dependent on rock type than climatic patterns that bring longshore currents and wave impact from a recurrent and propitious direction.

## 6. Conclusions

A multifaceted approach to our study of the natural harbor at Puerto Escondido and its development over time involved aspects of geology, geomorphology, paleontology, and sedimentology, leading to the following conclusions:

- During the last interglacial epoch near the close of Pleistocene time 125,000 years ago, the inner shores at Puerto Escondido were exposed to normal sea water that resulted in a substantial shell deposit dominated by a few marine invertebrates with a preference for intertidal conditions. By comparison, the present-day lagoon at the center of the inner harbor is far more restricted in its marine ecology. Therefore, the hills around the outer margin must have had gaps that permitted sufficient marine circulation to generate the Pleistocene shell deposit.

- The exaggerated topographic asymmetry expressed today by hills on the outer shores of Puerto Escondido was likely initiated through the coastal erosion along a major fault scarp shortly after the Late Pleistocene, but ongoing erosion during the Holocene provided more than enough raw materials to construct two natural barriers that closed off the large inner harbor. A parallel fault trace through the inner hills on the west side of the harbor accounts for post-depositional uplift of the Pleistocene shell beds.

- Shallow earthquakes are a common occurrence throughout the Gulf of California, but no credible evidence exists that tsunamis ever played a role in the geomorphology of coastal or basin sedimentology. Longshore currents related to strong seasonal winds are an important factor in the gulf's pattern of marine circulation and shore erosion on an annual basis. More common during El Niño years, hurricanes are the leading cause of rocky-shore erosion and development of coastal boulder deposits. Local production of boulders approaching a metric ton in weight can only be explained by sea surge stimulated by hurricane-force winds.

- Comparisons with other boulder deposits in the Gulf of California suggest that a named sub-category is appropriate to describe barriers or bars where large boulders are entrained in linear structures linked at one or both ends to sea cliffs. More often, a coastal boulder deposit (CBD) refers to extra-large clasts at the top of or at the immediate side of sea cliffs where natural jointing controls the initial size and shape of eroded materials. Herewith, we propose the term boulder barrier deposit (BBD) for geomorphic features that appear to be especially widespread around the Gulf of California. Future contributions are invited to explore the extent to which such features are well developed elsewhere around the world.

**Author Contributions:** Initial field reconnaissance was conducted by M.E.J. in February 2018, with a follow-up visit by M.E.J. and E.M.J. to collect field data in April 2019. M.E.J. prepared the first draft of this contribution, drafted all figures, and supplied all ground photos. R.G.-F. was responsible for working out the mathematics related to storm hydrodynamics. J L.-V. summarized the literature on basin turbidites in the lower Gulf of California that negates claims regarding tsunami activity in the region. In addition, J.L.-V. contributed input on fault orientations within the study area. Authorship has been limited to those who contributed substantially to the work reported. All authors have read and agreed to the published version of the manuscript.

**Funding:** This research received no external funding.

**Acknowledgments:** Foremost, we are grateful to Norm Christy, part-time resident of Loreto, for his crucial assistance with logistics during our 2019 visit including guidance on overland access to the boulder barriers at Puerto Escondido. Use of a vehicle and two-person kayak was made available for completion of fieldwork. M.E.J. is indebted to Jay Racela (Environmental Lab, Williams College) for help with the experimental calculation of density for the andesite sample from Puerto Escondido. Reviews of an earlier manuscript for which the authors are most grateful were provided by two anonymous readers.

**Conflicts of Interest:** The authors declare no conflict of interest.

## References


1. Backus, D.H.; Johnson, M.E.; Ledesma-Vazquez, J. Peninsular and island rocky shores in the Gulf of California. In *Atlas of Coastal Ecosystems in the Western Gulf of California*; Johnson, M.E., Ledesma-Vazquez, J., Eds.; University Arizona Press: Tucson, AZ, USA, 2009; pp. 11–27. ISBN 978-0-8165-2530-0.

2. Johnson, M.E.; Ledesma-Vázquez, J.; Guardado-France, R. Coastal geomorphology of a Holocene hurricane deposit on a Pleistocene marine terrace from Isla Carmen (Baja California Sur, Mexico). *J. Mar. Sci. Eng.* **2018**, *6*, 108. [CrossRef]

3. Johnson, M.E.; Guardado-France, R.; Johnsob, E.M.; Ledesma-Vázquez, J. Geomorphology of a Holocene Hurricane deposit eroded from rhyolite sea cliffs on Ensenada Almeja (Baja California Sur, Mexico). *J. Mar. Sci. Eng.* **2019**, *7*, 193. [CrossRef]

4. Steinbeck, J.; Ricketts, E.F. *Sea of Cortez: A Leisurely Journal of Travel and Research*; Viking Press: New York, NY, USA, 1941; p. 598.

5. Ricketts, E.F. *Between Pacific Tides*; University of Stanford Press: Palo Alto, CA, USA, 1939.

6. Merrifield, M.A.; Winant, C.D. Shelf-circulation in the Gulf of California: A description of the variability. *J. Geophy. Res.* **1989**, *94*, 133–160. [CrossRef]

7. Ledesma-Vázquez, J.; Johnson, M.E.; Gonzalez-Yajimovich, O.; Santamaría-del-Angel, E. Gulf of California geography, geological origins, oceanography and sedimentation patterns. In *Atlas of Coastal Ecosystems in the Western Gulf of California*; Johnson, M.E., Ledesma-Vázquez, J., Eds.; University of Arizona Press: Tucson, AZ, USA, 2009; pp. 1–10. ISBN 978-0-8165-2530-0.

8. Romero-Vadillo, E.; Zaystev, O.; Morales-Pérez, R. Tropical cyclone statistics in the northeastern Pacific. *Atmósfera* **2007**, *20*, 197–213.

9. Muriá-Vila, D.; Jaimes, M.Á.; Pozos-Estrada, A.; López, A.; Reinoso, E.; Chávez, M.M.; Peña, F.; Sánchez-Sesma, J.; López, O. Effects of hurricane Odile on the infrastructure of Baja California Sur, Mexico. *Nat. Hazards* **2018**, *9*, 963–981. [CrossRef]

10. Mark, C.; Gupta, S.; Carter, A.; Mark, D.F.; Gautheron, C.; Martin, A. Rift flank uplift at the Gulf of California: No requirement for asthenosperhic upwelling. *Geology* **2014**, *42*, 259–263. [CrossRef]

11. Carreño, A.L.; Helenes, J. Geology and ages of the islands, 14–40. In *A New Island Biogeography of the Sea of Cortés*; Case, T.J., Cody, M.L., Ezcurra, E., Eds.; Oxford University Press: Oxford, UK, 2002; p. 669.

12. Wentworth, C.K. A scale of grade and class terms for clastic sediments. *J. Geol.* **1922**, *27*, 377–392. [CrossRef]

13. Sneed, E.D.; Folk, R.L. Pebbles in the lower Colorado River of Texas: A study in particle morphogenesis. *J. Geol.* **1958**, *66*, 114–150. [CrossRef]

14. Nott, J. Waves, coastal bolder deposits and the importance of pre-transport setting. *Earth Planet. Sci. Lett.* **2003**, *210*, 269–276. [CrossRef]

15. Orlieb, L.; Dauphin, J.P. *Quaternary Vertical Movements along the Coasts of Baja California and Sonora*; Dauphin, J.P., Simoneit, B.R.T., Eds.; American Association of Petroleum Geologists Memoir 47: Gulf, CA, USA; Peninsular, CA, USA, 1991; pp. 447–480.

16. Bender, M.L.; Fairbanks, R.G.; Taylor, F.W.; Matthews, R.K.; Goddard, J.G.; Broecker, W.S. Uranium-series dating of the Pleistocene reef tracts of Barbados, West Indies. *Geol. Soc. Am. Bull.* **1979**, *90*, 577–594. [CrossRef]
17. Neumann, A.C.; Hearty, P.J. Rapid sea-level changes at the close of the last interglacial (substage 532) recorded in Bahamian island geology. *Geology* **1996**, *24*, 775–778. [CrossRef]
18. Brusca, R.C. *Common Intertidal Invertebrates of the Gulf of California*, 2nd ed.; Universityu of Arizona Press: Tucson, AZ, USA, 1973; p. 513.
19. Hayes, M.L.; Johnson, M.E.; Fox, W.T. Rocky-shore biotic associations and their fossilization potential: Isla Requeson (Baja California Sur, Mexico). *J. Coast. Res.* **1993**, *9*, 944–957.
20. Simian, M.E.; Johnson, M.E. Development and foundering of the Pliocene Santa Ines Archipelago in the Gulf of California: Baja California Sur, Mexico. In *Pliocene Carbonates and Related Facies Flanking the Gulf of California, Baja California Mexico, Geological Society of America Special Paper*; Geological Society of America: McLean, VA, USA, 1997; Volume 318, pp. 25–38.
21. McCloskey, T.A.; Bianchette, T.A.; Liu, K.-B. Geological and sedimentological evidence of a large tsunami occurring ~1100 year BP from a small coastal lake along the Bay of La Paz in Baja California Sur, Mexico. *J. Mar. Sci. Eng.* **2015**, *3*, 1544–1567. [CrossRef]
22. Gorsline, D.S.; De Diego, T.; Nava-Sanchez, E.H. Seismically triggered turbidites in small margin basins: Alfonso Basin, Western Gulf of California and Santa Monica Basin, California Borderland. *Sediment. Geol.* **2000**, *135*, 21–35. [CrossRef]
23. Castro, R.; Stock, J.M.; Hauksson, E.; Clayton, R.W. Active tectonics in the Gulf of California and seismicity (M > 3.0) for the period 2002–2014. *Tectonophysics* **2017**, *719–720*, 4–16. [CrossRef]
24. Fletcher, J.M.; Munguía, L. Active continental rifting in southern Baja California, Mexico: Implications for plate motion partitioning and the transition to seafloor spreading in the Gulf of California. *Tectonics* **2000**, *19*, 1107–1123. [CrossRef]
25. Umhoefer, P.J.; Mayer, L.; Dorsey, R.J. Evolution of the margin of the Gulf of California near Loreto, Baja California Peninsula, Mexico. *Geol. Soc. Am. Bull.* **2002**, *114*, 849–868. [CrossRef]
26. Trejo-Gómez, E.; Ortiz, M.; Núñez-Cornú, J. Source model of the October 9, 1995 Jalisco-Colima tsunami as constrained by field survey reports and on the numerical simulation of the tsunami. *Geofís. Int.* **2015**, *54*, 149–159. [CrossRef]
27. Suanez, S.; Fichaut, B.; Magne, R. Cliff-top storm deposits on Banneg Island, Briggany, Rrance: Effects of giant waves in the eastern Atlantic Ocean. *Sediment. Geol.* **2009**, *220*, 12–28. [CrossRef]
28. Cox, R.; Jahn, K.L.; Watkins, O.G.; Cox, P. Extraordinary boulder transport by storm waves (west Ireland, winter 2013-14) and criteria for analyzing coastal boulder deposits. *Earth Sci. Rev.* **2018**, *177*, 623–636. [CrossRef]
29. Erdmann, W.; Scheffers, A.M.; Kelletat, D.H. Holocene coastal sedimentation in a rocky environment: Geomorphological evidence from the Aran Islands and Galway Bay (Western Ireland). *J. Coast. Res.* **2018**, *34*, 772–792. [CrossRef]
30. Hearty, P.J. Boulder deposits from large waves during the last inter-glaciation on North Eleuthera Island, Bahamas. *Quat. Res.* **1997**, *48*, 326–338. [CrossRef]
31. Rovere, A.; Casella, E.; Harris, D.L.; Lorscheid, T.; Nandasena, N.A.K.; Dyer, B.; Sandstrom, M.R.; Stocchi, P.; D'Amdrea, W.J.; Raymo, M.E. Giant boulders and last interglacial storm intensity in the North Atlantic. *Proc. Natl. Acad. Sci. USA* **2017**, *114*, 12144–12149. [CrossRef] [PubMed]
32. Kennedy, A.B.; Mori, N.; Yasuda, T.; Shimozono, T.; Tomiczek, T.; Donahue, A.; Shimura, T.; Imai, Y. Extreme block and boulder transport along a cliffed coastline (Calicoan Island, Philippines) during Super Typhoon Haiyan. *Mar. Geol.* **2017**, *383*, 65–77. [CrossRef]
33. Biolchi, S.; Denamiek, C.; Devoto, S.; Korbar, T.; Macovaz, V.; Scicchitano, G.; Vilibic, I.; Furlani, S. Impact of the October 2018 Storm Vaia on coastal boulders in the northern Adriatic Sea. *Water* **2019**, *11*, 2229. [CrossRef]

34. Johnson, M.E.; Ledesma-Vázquez, J.; Backus, D.H.; González, M.R. Lagoon microbialites on Isla Angel de la Guarda and associated peninsular shores, Gulf of California (Mexico). *Sediment. Geol.* **2012**, *263*, 76–84. [CrossRef]
35. Johnson, M.E.; Ledesma-Vázquez, J. Biological zonation on a rocky-shore boulder deposit: Upper Pleistocene Bahía San Antonio (Baja California Sur, Mexico). *Palaios* **1999**, *14*, 569–584. [CrossRef]

Journal of
*Marine Science and Engineering*

*Article*

# Upper Pleistocene and Holocene Storm Deposits Eroded from the Granodiorite Coast on Isla San Diego (Baja California Sur, Mexico)

Ginni Callahan [1], Markes E. Johnson [2,*], Rigoberto Guardado-France [3] and Jorge Ledesma-Vázquez [3]

[1]  Sea Kayak Baja Mexico, Nicolas Bravo s/n, Entre Baja California y El Muro de Contención, Colonia Centro, Loreto 23880, Baja California Sur, Mexico; ginni@seakayakbaja.com
[2]  Department of Geosciences, Williams College, Williamstown, MA 01267, USA
[3]  Facultad de Ciencias Marinas, Universidad Autónoma de Baja California, Ensenada 22800, Baja California, Mexico; rigoberto@uabc.edu.mx (R.G.-F.); ledesma@uabc.edu.mx (J.L.-V.)
*  Correspondence: markes.e.johnson@williams.edu; Tel.: +1-413-2329

**Abstract:** This project examines the role of hurricane-strength events likely to have exceeded 119 km/h in wind speed that entered the Gulf of California from the open Pacific Ocean during Late Pleistocene and Holocene times to impact the granodiorite shoreline on Isla San Diego. Conglomerate dominated by large, ellipsoidal to subspherical boulders at the islands south end were canvassed at six stations. A total of 200 individual cobbles and boulders were systematically measured in three dimensions, providing the database for analyses of variations in clast shape and size. The project's goal was to apply mathematical equations elaborated after Nott (2003) with subsequent refinements to estimate individual wave heights necessary to lift igneous blocks from the joint-bound and exfoliated coast on Isla San Diego. On average, wave heights on the order of 3 m are calculated as having impacted the Late Pleistocene rocky coastline on Isla San Diego during storms, although the largest boulders more than a meter in diameter are estimated to weigh two metric tons and would have required waves in excess of 10 m for extraction. Described for the first time, a fossil marine biota associated with the boulder beds confirms a littoral-to-very-shallow water setting correlated with Marine Isotope Substage 5e approximately 125,000 years ago. A narrow submarine ridge consisting, in part, of loose cobbles and boulders extends for 1.4 km to the southwest from the island's tip, suggesting that Holocene storms continued to transport rock debris removed from the shore. The historical record of events registered on the Saffir–Simpson Hurricane Wind Scale in the Gulf of California suggests that major storms with the same intensity struck the island in earlier times.

**Keywords:** coastal erosion; storm surge; hydrodynamic equations; Marine Isotope Substage 5e; Gulf of California

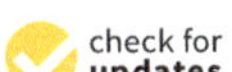

**Citation:** Callahan, G.; Johnson, M.E.; Guardado-France, R.; Ledesma-Vázquez, J. Upper Pleistocene and Holocene Storm Deposits Eroded from the Granodiorite Coast on Isla San Diego (Baja California Sur, Mexico). *J. Mar. Sci. Eng.* **2021**, *9*, 555. https://doi.org/10.3390/jmse9050555

Academic Editor: Matthew Lewis

Received: 19 April 2021
Accepted: 19 May 2021
Published: 20 May 2021

**Publisher's Note:** MDPI stays neutral with regard to jurisdictional claims in published maps and institutional affiliations.

## 1. Introduction

Oriented northeast to southwest between mainland Mexico and the Baja California Peninsula, 40 named islands in the Gulf of California spread out over a sea surface of 160,000 km$^2$. These gulf islands range between 1224 km$^2$ and 22 ha in size [1]. Island development postdates the opening of the gulf to the Pacific Ocean by rifting from the mainland more than 5 million years ago and many formed as fault blocks influenced by regional tectonics. Most are composed of Miocene volcanic flows or from intrusive igneous rocks of yet older Cretaceous origin. Of the 40 islands, 8 islands fall into the category dominated by granite or closely related granodiorite, and this study looks at one of the smallest in the lower Gulf of California called Isla San Diego with an area of 60 ha [1]. Survey work conducted through satellite imagery shows that rocky shores account for nearly half the gulf's peninsular coastline including related islands [2]. At slightly more than 23%, andesite dominates the region's total shores, followed by granite



*J. Mar. Sci. Eng.* **2021**, *9*, 555. https://doi.org/10.3390/jmse9050555

https://www.mdpi.com/journal/jmse

or granodiorite at 9%, and limestone at 7.5%, while rock types including other igneous rocks or metamorphic rocks are less well represented.

This contribution belongs to a series of papers focused on the erosion of coastal boulder beds from their parent rocks within the Gulf of California. Upper Pleistocene and Holocene deposits formed by boulders are commonly found along the peninsular shores of Baja California and around the gulf islands but most studies in coastal geomorphology seldom compare the results of rock density on an interregional basis as related to different parent rock types. The application of mathematical formulae to estimate storm wave height was applied previously to coastal boulder deposits throughout the Gulf of California, including those formed by limestone, rhyolite, and andesite clasts [3–6]. Extension of this program now includes the Pleistocene boulder beds eroded from the granodiorite coast of Isla San Diego, applying the same methodology of systematic size measurements to calculate volume and weight based on rock density preliminary to the estimation of wave heights derived from competing equations. The study also newly describes marine fossils preserved within the Pleistocene conglomerate of Isla San Diego that date the deposits with reasonable accuracy. Finally, development during the Holocene time of a long marine ridge off the southwestern tip of the island brings into consideration the ongoing influence of hurricanes capable of moving large boulders in a shallow, subtidal setting.

Aside from the limited statistics available on the size, geologic origins, and coastal composition of islands in the Gulf of California [1,2], barely any literature exists on the geology and geomorphology of Isla San Diego except for an early nineteenth-century appraisal that includes the only previous description of the submarine ridge off the island's southernmost end [7]. Attention to the phenomenon of coastal mega boulders and their relationship to major storms or tsunami events is a topic of growing interest [8–11]. Especially in the context of rock density, the data from Isla San Diego provide further insight on comparison of storm beds of Pleistocene and Holocene origins throughout the gulf region [3–6] with oceanic basalt in the Azores and Canary Islands of the North Atlantic [12,13], as well as rare mantel rocks from storm beds in coastal Norway on the Norwegian Sea [14].

## 2. Geographical and Geological Setting

Stretching for more than 1000 km in length (Figure 1a), the Gulf of California is a marginal sea seated over a tectonically active zone that entails spreading centers offset by a succession of transform faults [15]. The central spine of the adjacent Baja California Peninsula is formed by granodiorite, broadly dated to a Cretaceous origin between 97 and 90 million years ago [16]. Upfaulted granodiorite basement occurs on the Baja California Peninsula at Punta San Antonio north of Loreto, on Isla Catalina east of Loreto, as well as Isla Santa Cruz and Isla San Diego (Figure 1a). Peninsular and island development including those areas with granodiorite resulted from Miocene extensional rifting that began prior to flooding 13 million years ago and lasted for 9.5 million years when a change in dynamics initiated transform faults connected with the San Andreas Fault on the US side of the border [1,2,15]. A detachment zone was activated approximately 3.5 million years ago that resulted in half-graben structures separating the islands from the rest of the Baja California Peninsula. The detachment zone that extends from Punta San Antonio to Isla San Diego (Figure 1a) is identified as the Comondú Detachment.

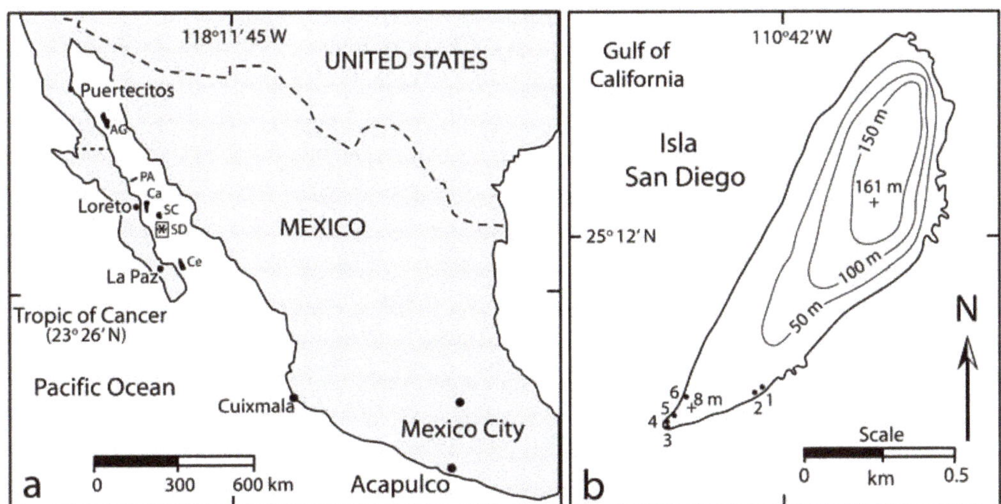



**Figure 1.** Mexico's Baja California Peninsula and Isla San Diego: (**a**) map showing the boundary between the United States and Mexico as well as the boundary between the northern and southern states of Baja California and Baja California Sur (dashed lines), together with towns on the Baja California Peninsula and key spots or islands including Ángel de la Guarda (AG), Punta San Antonio (PA), Carmen (Ca), Santa Catalina (SC), Cerralvo (Ce) and San Diego (box with asterisk); (**b**) enlarged map of Isla San Diego in the lower Gulf of California, showing the location of Stations 1 to 6 where cobbles and boulders of eroded granodiorite were measured for this study.

Isla San Diego is at the southeast end of the detachment zone 20 km or 11 nautical miles due east of the closest access point on the Baja California Peninsula. The main north–south highway is too distant from the peninsula's eastern shore to make boat access to the island convenient. Compared to other islands farther to the northwest or southeast, the relative isolation of Isla San Diego meant that it received little attention from geographers and geologists. No formal topographic map by the Federal Mexican government exists for Isla San Diego and its dimensions and topography were appraised by satellite imagery [2]. The island is elongated in shape, approximately 1.5 km in length and 0.43 km in width with northeast to southwest orientation (Figure 1b). The maximum elevation is more than 160 m above mean sea level, as attained to the north, but the island's central ridgeline tapers gradually downward to the shore at the southwest end.

The island core is composed entirely of granodiorite and the conglomerate beds eroded from these basement rocks occur exclusively at the southwestern end. A prominent submarine ridge extends from the tip of the island [7], where large boulders of loosely piled granodiorite are close to the surface (Figure 2a). Granodiorite sea cliffs are well exposed along the east shore for more than 400 m from the southwest tip of the island to the northeast, where a series of closely spaced grottos are eroded as a result of spheroidal weathering between joints in the rock (Figure 2b). Additional weathering along both flanks of the island is the result of sheeted exfoliation typical of granitoid rocks.

The greater part of Mexico's Natural Protected Areas (Áreas Naturales Protegida) is taken up by the Gulf of California Biosphere Reserve protecting all islands in the Gulf of California, which accounts for roughly 19% of the nation's total conservation reserves [17]. Therefore, Isla San Diego is protected under conservation guidelines due to its biodiversity and ecological characteristics. All materials and fossils identified in this study were left in place on Isla San Diego.

**Figure 2.** South end of Isla San Diego (lower Gulf of California: (**a**) view showing part of the shallow-water ridge composed of loose cobbles and boulders of eroded granodiorite oriented S 55° W off the island; (**b**) southwest end of the island showing small sea caves eroded in granodiorite basement rocks overlain by Pleistocene conglomerate.

### 3. Materials and Methods

*3.1. Data Collection*

The raw data for this study were collected in March 2021 from deposits composed exclusively of granodiorite cobbles and boulders consolidated by a thin limestone matrix. Individual clasts from six stations were measured manually to the nearest half centimeter in three dimensions perpendicular to one another (long, intermediate, and short axes). Differentiated from cobbles, the base definition for a boulder adapted in this exercise was that of Wentworth [18] for an erosional clast equal or greater than 25.6 cm in diameter. Triangular plots were employed to show variations in clast shape, following the design of Sneed and Folk [19] for river pebbles. In the field, all measured clasts were characterized as subrounded, and a smoothing factor of 20% was applied uniformity to adjust for the estimated volume calculated by the simple multiplication of length from the three axes. Comparative data on maximum cobble and boulder dimensions were fitted to bar graphs to show size variations in the long and intermediate axes from one sample to the next. The rock density from a granodiorite sample yielded a value of 2.52 g/cm$^3$.

*3.2. Hydraulic Model*

Granodiorite is the typical intrusive magmatic rock characteristic of several islands in the Gulf of California. Herein, two formulas were applied to estimate the size of storm waves against joint-bound blocks. Equation (1) derives from the work of Nott [20] and Equation (2) is modified from an alternative approach using the velocity equations of Nandasena et al. [21] applied to storm deposits by Pepe et al. [22].

$$Hs = \frac{\left(\frac{\rho_s - \rho_w}{\rho_w}\right)a}{C_1} \qquad (1)$$

$$Hs = \frac{2\left(\frac{\rho_s - \rho_w}{\rho_w}\right) \cdot c \cdot [\cos\theta + (\mu_s \cdot \sin\theta)]}{c_1}\Bigg/100 \tag{2}$$

where $Hs$ = height of the storm wave at breaking point; $\rho_s$ = density of the boulder (tons/m$^3$ or g/cm$^3$); $\rho_w$ = density of water at 1.02 g/mL; $a$ = length of the boulder on long axis in cm; $\theta$ is the angle of the bed slope at the pretransport location (1° for joint-bounded boulders); $\mu_s$ is the coefficient of static friction (=0.7); and $C_1$ is the lift coefficient (=0.178). Equation (1) is more sensitive to the length of a boulder on the long axis, whereas Equation (2) is more sensitive to the length of a boulder on the short axis. Therefore, some differences are expected in the estimates of $Hs$.

## 4. Results

### 4.1. Base Map and Sample Stations

Isla San Diego is among the smallest named islands formed by granodiorite in the Gulf of California [1]. Its size was conducive to a close coastal survey by kayak that allowed for the location and appraisal of conglomerate beds on the island's periphery. Six sampling stations were chosen from the conglomerate outcrops found only around the southwestern end of the island (Figure 1b). Between 30 and 35 individual clasts were measured within a meter's radius from a position above the source basement rock. Co-ordinates are listed in Appendix A (Tables A1–A6) for each station recorded by a hand-held device for tracking by the satellite-based global positioning system (GPS). Sample Stations 1 and 2 (Figure 3a) are located on the east shore 375 m and 350 m north of the island's tip, respectively, where the conglomerate sits directly above a bench of granodiorite approximately 1.5 m above mean sea level. Station 3 is located at the extreme southwestern tip of the island (Figure 3b), where crude layering in the conglomerate shows a 30° inclination to the northwest. Three additional stations were established on the west shore (Figure 1b), where the contact is concealed by talus. Clasts measured at those stations also were limited to a 2 m radius at points near the bottom of the conglomerate bed but included some samples from the talus showing evidence of carbonate cement formerly binding the conglomerate. Clean clasts from the intertidal zone on that side of the island are reworked by coastal currents and were excluded.

**Figure 3.** Sample stations on the southeast side and southern tip of Isla San Diego: (**a**) sample Stations 1 and 2 occur at 375 m and 350 m north of the island's southern tip (red circles are 2 m in diameter); (**b**) sample Station 3 is located at the far southwestern tip of Isla San Diego (red oval is 0.5 m wide).

### 4.2. Comparative Variation in Clast Shapes

Raw data on clast size in three dimensions collected from each of the six sampling stations are recorded in Appendix A (Tables A1–A6). With regard to shape, points representing individual cobbles and boulders were fitted to a set of Sneed–Folk triangular diagrams (Figure 4a–f). The slope of points is in general agreement among the six plots, following a uniformly diagonal trend from the middle of the second tier to the lower right-hand rhomboid. Erosional wear on a perfect cube at all four corners results in a clast with equal values in three dimensions that will plot at the apex of the small triangle in the topmost tier. Variations that reflect slightly smaller values for the intermediate and short axis will shift the location more toward the center of that space. Only one or two clasts fall into this field, which indicates that vertical joints and horizontal fractures in the parent granodiorite are not evenly spaced in an orderly three-dimensional grid. Any point that falls into the center of the rhomboid on the right-hand end of the lower tier represents an individual clast with a long axis twice the length of the intermediate axis perpendicular to it, which, in turn, measures five times the length of the short axis perpendicular to the other two. The form of such a clast is initially bar-shaped but becomes more spindle-shaped as the sharp edges at the corners are worn away by abrasion.

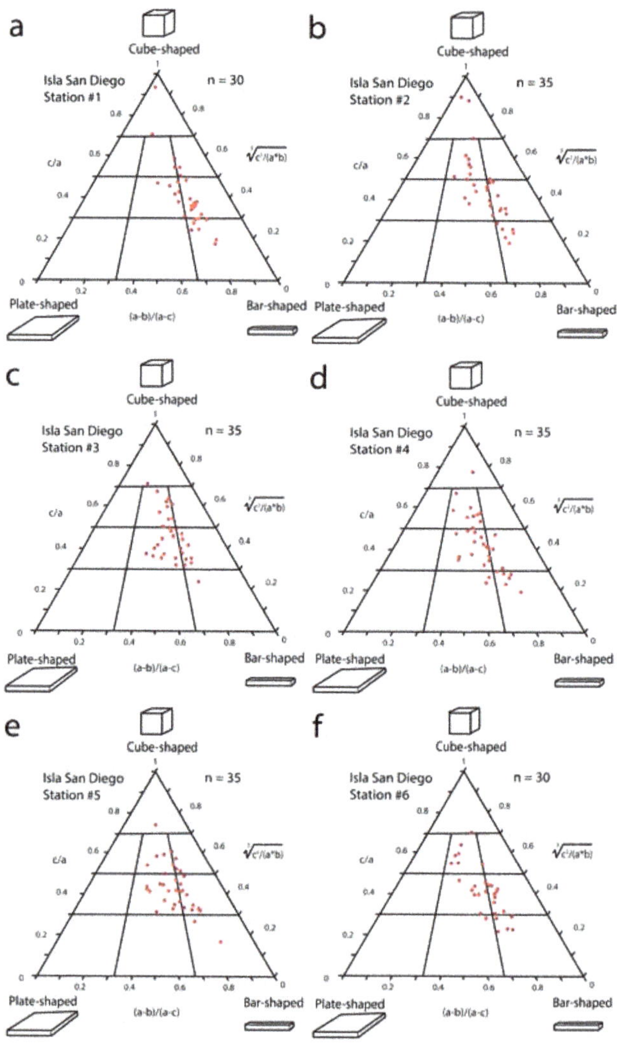


**Figure 4.** Set of triangular Sneed–Folk diagrams used to appraise variations in cobble and boulder shapes sampled along the Upper Pleistocene paleoshore on Isla San Diego in the lower Gulf of California ((**a–f**), Stations 1–6).

The greatest number of points from each of the six samples falls within the third tier from the top but on opposite sides of the line separating the right side of the diagram from the center. A particular clast with dimensions in which the intermediate and short axis are close in value, but roughly half that of the long axis will fall squarely onto the midline between the two rhomboids at the center of the diagram. As the third dimension (shortest axis) decreases in length, the point will shift in position across the line to the right and lower downward in position. Overall, the comparative results of shape analysis indicate that most of the cobbles and boulders in the Pleistocene conglomerate are elongated in shape but as relatively fat spindles with an ellipsoidal outline. It is important to distinguish overall size from shape. That is to say, a smaller cobble with the same ratio of measurements between long, intermediate, and short axes will plot exactly the same as a larger boulder with the same ratios. Although clasts were chosen at random, a reasonably large population of clasts within a limited search radius at any one station assures that the sample is representative. In this case, the absence of more perfectly spherical clasts, and the dominant trend toward thickened spindles is evident.

### 4.3. Comparative Variation in Clast Sizes

Drawn from original data (Tables A1–A6), clast size is treated separately to best effect on bar graphs as a function of frequency against maximum and intermediate lengths of the two longest axes perpendicular to one another. The first set of six graphs so plotted (Figure 5a–f) exhibit trends in the maximum dimension for clast length sorted by intervals of 15 cm in which the boundary between cobbles and boulders is marked within the range for clasts between 16 and 30 cm in diameter. The pattern is compared with another six graphs (Figure 6a–f) based on measurements for the length of the intermediate axis in the same 200 clasts.

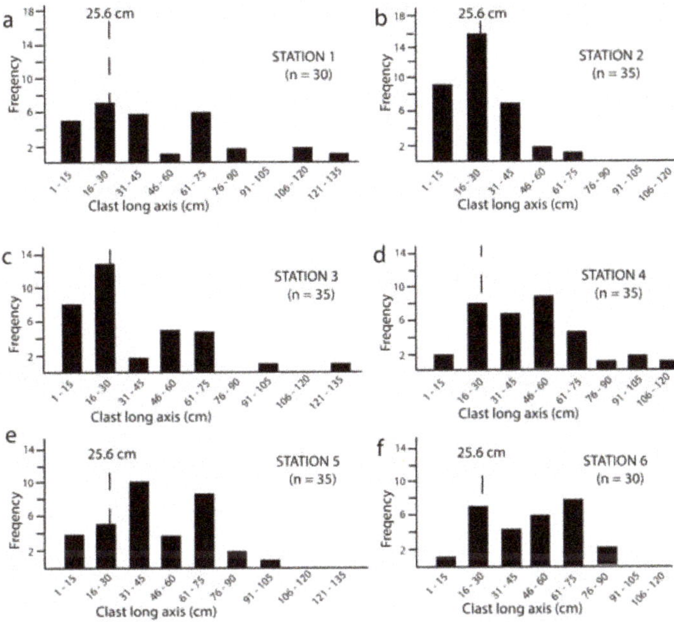

**Figure 5.** Set of bar graphs used to contrast variations in the maximum size of clast axes from six samples at Isla San Diego in the lower Gulf of California: (**a**) bar graphs from Station 1; (**b**) bar graphs from Station 2; (**c**) bar graphs from Station 3; (**d**) bar graphs from Station 4; (**e**) bar graphs from Station 5; (**f**) bar graphs from Station 6. Dashed line (offset to represent 26.6 mm) marks the boundary between large cobbles and small boulders.

Based on four of the six graphs (Figure 5a,d–f) representing maximum axial length, it is shown that boulders outnumber the smaller cobbles at rations between 3:1 and 3:2. The opposite is indicated by two of the graphs (Figure 5b,c) in which the smaller cobbles outnumber the larger boulders at ratios 2:1 or less. Based on two of the six graphs (Figure 6b,c) representing the intermediate axial length for the same 200 clasts, it may be argued that cobbles outnumber the larger boulders in only two of the graphs (Figure 6b,c), whereas cobbles are slightly outnumbered by boulders in two graphs (Figure 6a,d) and occur at parity in two others (Figure 6a,f). Overall, comparative data between maximum length and intermediate axial length confirm the results on clast shape (Figure 4), showing the dominance of ellipsoidal boulder shapes.

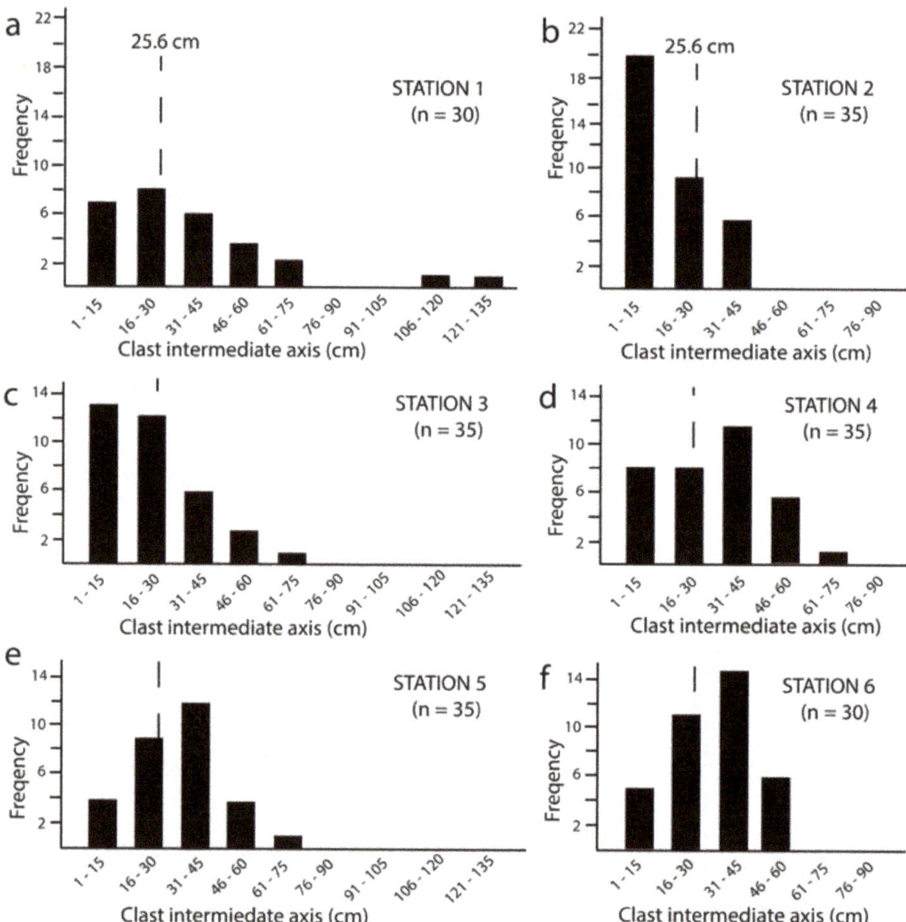


**Figure 6.** Set of bar graphs used to contrast variations in the intermediate size of clast axes from six samples at Isla San Diego in the lower Gulf of California: (**a**) bar graphs from Station 1; (**b**) bar graphs from Station 2; (**c**) bar graphs from Station 3; (**d**) bar graphs from Station 4; (**e**) bar graphs from Station 5; (**f**) bar graphs from Station 6. Dashed line (offset to represent 26.6 mm) marks the boundary between large cobbles and small boulders.

### 4.4. Clast Imbrication

Clast orientation or imbrication within the boulder deposit on Isla San Diego varies according to outcrop exposure, as observed on two vertical planes that intersect perpendicular to one another. One is parallel to the island's long axis along the eastern shore (Figures 2b and 3a). The other crosses through the island's southern tip (Figure 3b). Both show direct contact of eroded boulders with underlying granodiorite basement rocks where the conglomerate attains a maximum elevation of 8 m above sea level. The unconformity surface traced parallel to the island's axis is flat lying with a slight gain in elevation rising to the northeast. In this view (Figure 2b), a change in clast size from boulders to large cobbles begins above a reactivation surface that follows a minor indentation in the cliffs obscured by shadows. Shape analyses indicate that a preponderance of clasts from the basal part of the deposit are ellipsoidal or roughly fusiform in shape (Figure 4), but a small amount of imbrication is detected only below the reactivation surface (Figure 2b). The unconformity surface exposed in the plane crossing the tip of the island (Figure 3b) is more irregular with a dip or swale in the center, but crude layering in the overlying boulder deposit is increasingly inclined with distance above the unconformity. Based on photographic evidence supplemental to Figure 3b, evidence for imbrication is detected in the upper part of the deposit (Figure 7), where this is shown by transfer of outline tracings (Figure 7a,b) and isolated for clearer viewing (Figure 7c). Due to the partial collapse of clasts at the island's tip, relationships among those in the basal part of the deposit are less clear. However, the orientation of clasts from the upper part of the deposit (Figure 7c) reveals a pattern of imbrication from northeast to southwest.

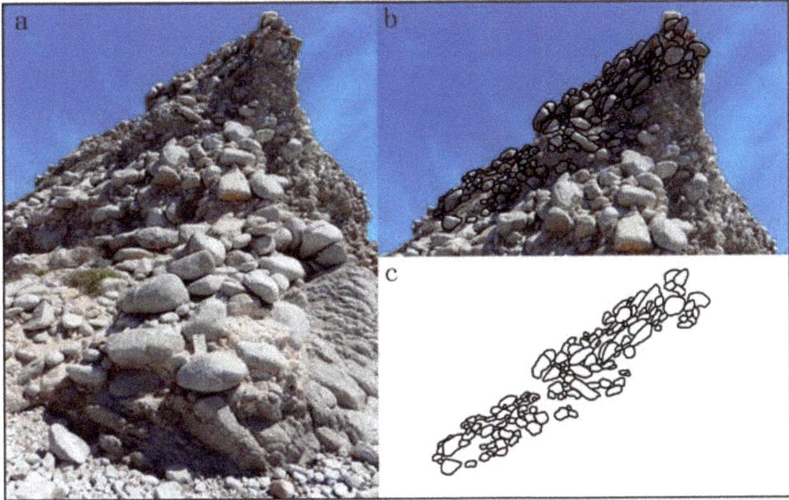


**Figure 7.** Photographic evidence for clast imbrication: (**a**) view overlooking sample Station 3 at the southern tip of Isla San Diego; (**b**) upper part of the conglomerate deposit at the same locality, tracing the outline of boulders; (**c**) drawing of the tracing isolated for clarity.

### 4.5. Fossil Fauna with Inferences on Age and Water Depth

A mixed fauna of three fossil corals, four bivalves, a single gastropod, and a coral-boring barnacle is preserved among the granodiorite cobbles and boulders on Isla San Diego (Table 1). None of the fossils occur as encrustations attached to individual cobbles or boulders. With possible exceptions among certain bivalves, they qualify as organic clasts that became secondarily incorporated within the conglomerate.

**Table 1.** Summary list of marine invertebrate fossils from the Upper Pleistocene boulder beds on Isla San Diego correlated with Marine Isotope Substage 5e.

| Phylum | Class | Species | Phylum | Class | Species |
|---|---|---|---|---|---|
| Coelenterata | Anthozoa | *Porites panamensis*<br>*Pocillopora elegans*<br>*Povona gigantea* | Mollusca | Bivalvia | *Codakia distinguenda*<br>*Lyropecten subnodosus*<br>*Ostrea* sp.<br>*Spondylus calcifer* |
| | | | Arthropoda | Gastropoda<br>Cirripedia | *Turbo fluctuosus*<br>*Hexacreusia durhami* |

Representative fossils are illustrated by field photos (Figure 8). Two kinds of corals are illustrated: *Porites panamensis* (Figure 8a) and *Pocillopora elegans* (Figure 8b), respectively. Articulated bivalves are rarely found as fossils within the boulder beds, with the exception of *Codakia distinguenda* (Figure 8c), which may have grown in place in cavities among boulders after their deposition. Very thick but heavily eroded shell fragments belonging to a species of oyster (Figure 8d) indicate that shell fragmentation normally occurred prior to burial in the conglomerate. The large and heavily calcified shell of *Spondylus calcifer* (Figure 8e) is preserved intact but disarticulated. The large pecten *Lyropecten subnodosus* (Figure 8f) is likewise disarticulated and also broken. The only trace found of fossil gastropods is the distinctive operculum belonging to *Turbo fluctuosus*. Of ecological note, one of the *Porites* fossils observed in the boulder deposit at Station 4 (Figure 7a) is host to the boring barnacle *Hesareusia durhami* (Figure 8g). The same relationship between coral host and barnacle is known from the Pleistocene reef complex preserved intact on Isla Cerralvo [23].

**Figure 8.** Upper Pleistocene fossils from the granodiorite conglomerate at Isla San Diego: (**a**) filling among boulders that includes the coral *Porites panamensis* (ruler for scale); (**b**) broken branch from the coral *Pocillopora elegans* (approximately 3 cm in diameter); (**c**) articulated bivalve (*Codakia distinguenda*) with ruler for scale; (**d**) fragment of a thick-shelled oyster (pen tip for scale;) (**e**) inside surface of a disarticulated valve belonging to *Spondylus calcifer* (approximately 7 cm in width); (**f**) broken shell belonging to the species *Lyropecten subnodosus* (ruler for scale); (**g**) detail from Figure 7a showing the surface opening of the barnacle *Hexacreusia durhami* (approximately 3 mm in diameter).

Fossil mollusks from the Gulf of California occupy a wide range of geological ages through the Pliocene and Pleistocene, but the corals are more nuanced. In a detailed survey of localities throughout Baja California and associated gulf islands [24], 35 out of 47 collection sites include the species *Porities panamensis*, most of which are dated as Late Pleistocene in age. *Pocillopira elegans* is reported previously from only a single locality, dated as Late Pleistocene [24]. A definitive age determination requires radiometric testing for lead isotopes not within the scope of this paper. Uncertain assignments of *P. panamensis* to a mid-Pleistocene age and older Pliocene times are limited to marine terraces or other inland localities well elevated by tectonic uplift above present-day sea level. The absence of marine terraces on Isla San Diego supports a Late Pleistocene age for the fossil fauna. Alone, the fossil corals denote shallow-water conditions, but the related fossil mollusks add additional support of an intertidal to the shallow subtidal origin of the mixed fauna that lived nearby.

Global sea level during the last Pleistocene interglacial period (Marine Isotope Substage 5e) was at its highest, calculated to have stood between 4 m and 6 m higher than today on the basis of changes in oxygen isotopes of planktonic foraminifera and other criteria [25,26]. In that case, the granodiorite rocks at the south end of Isla San Diego presently exposed above sea level would have been submerged and development of the overlying boulder deposit would have occurred in very shallow water, where an infusion of a thin limestone binding matrix insured stabilization.

### 4.6. Storm Intensity as Function of Estimated Wave Height

Clast sizes and maximum boulder volumes drawn from the six sample stations are summarized in Table 2, allowing for direct comparison of average values for all clasts, as well as values for the largest clasts in each sample based on Equations (1) and (2) derived from the work of Nott [20] and Pepe et al. [22].

**Table 2.** Summary data from Appendix A (Tables A1–A6) showing maximum bolder size and estimated weight compared to the average values for sampled boulders from each of the transects together with calculated values for wave heights estimated as necessary for boulder–beach mobility. Abbreviations: EAWH = estimated average wave height, EMWH = estimated maximum wave height.

| San Diego Station | Number of Samples | Average Boulder Volume (cm$^3$) | Average Boulder Weight (kg) | EAWH (m) Nott [20] | EAWH (m) Pepe et al. [22] | Max. Boulder Volume (cm$^3$) | Max. Boulder Weight (kg) | EMWH (m) Nott [20] | EMWH (m) Pepe et al. [22] |
|---|---|---|---|---|---|---|---|---|---|
| 1 | 30 | 89,079 | 224 | 3.2 | 3.0 | 780,800 | 1968 | 10.6 | 11.4 |
| 2 | 35 | 7763 | 33.8 | 1.9 | 1.9 | 84,816 | 214 | 5.1 | 7.1 |
| 3 | 35 | 29,884 | 75.7 | 2.8 | 2.8 | 299,850 | 756 | 10.4 | 9.3 |
| 4 | 35 | 39,007 | 98 | 3.7 | 3.4 | 263,516 | 664 | 9.3 | 9.9 |
| 5 | 35 | 33,456 | 83.4 | 3.6 | 3.3 | 157,303 | 396 | 8.3 | 8.0 |
| 6 | 30 | 33,688 | 85 | 3.2 | 2.9 | 93,960 | 238 | 6.2 | 5.0 |
| Average | 33.33 | 66,536 | 100 | 3.1 | 2.9 | 280,041 | 706 | 8.0 | 8.5 |

The Nott formula [20], provided in Equation (1), yields an average wave height of 3.1 m for the extraction of joint-bound blocks from granodiorite sea cliffs exposed at Isla San Diego, as tabulated for sample stations 1 to 6. A much larger value for a wave height of 8.0 m is calculated from the average of the largest single blocks of granodiorite recorded from the six stations based on the application of the same equation. The more sensitive to clast length from the short axis, the more sophisticated Equation (2) applied by Pepe et al. [22] yields values that are slightly lower for the estimated average wave height with a difference of 30 cm. On the other hand, the application of Equation (2) yields a higher value by a half meter for the average of the largest boulders, compared to that of Equation (1). Notably, the value for the maximum wave height for the six largest boulders based on Equation (1) is 2.5 times higher than the computed average for all 200 clasts. The difference is nearly 3.0 times greater comparing the maximum wave height for the same six boulders with the computed average for all 200 clasts based on Equation (2). Clearly, the pressure of extreme wave impact against the shore is necessary to loosen and dislodge the largest joint-bound blocks of granodiorite preserved in the Pleistocene cliff line on Isla San Diego, as characterized by the enormous block at Station 1 estimated to weigh nearly two metric tons (Table 1, Table A1)

*4.7. Holocene Ridge Offshore Isla San Diego*

A distinct feature in the geomorphology of Isla San Diego is the pointed beach at the island's southwestern tip that extends offshore along an underwater ridge (Figure 9). Aerial photos reveal shallow, aquamarine waters above the ridge, indicating its extension for a distance of 1.4 km on a compass heading of 224 degrees (S 55° W). Exploration by kayak over the first 250 m offshore confirms that the ridge is formed by loose cobbles and large boulders. Observed at different times under different sea conditions, it is notable that the smaller clasts shift in location, whereas the large boulders remain fixed on the ridge. How far and to what depth the boulder train extends before changing to gravel and sand toward the distal end is unknown.

**Figure 9.** View in the shadow of cliff-forming granodiorite boulders at the tip of Isla San Diego looking to the southwest across the beach and major extension of a submarine ridge composed of loose cobbles and boulders (see also Figure 2a viewed from the opposite direction).

## 5. Discussion

*5.1. Inclined Boulder Beds and Imbrication Pattern as Mitigating Factors*

Conglomerate beds at the southern terminus of Isla San Diego (Figure 3b) suggest crude layering with a 30° dip to the northwest. The contact with underlying granodiorite makes it difficult to know for certain if the entire island has been tilted due to tectonics or if the fracture pattern in the granodiorite conforms to a pattern with horizontal fractures at right angles to vertical joints, as implied by the small sea caves or grottos eroded under present-day conditions along the island's eastern shore. The more complicated Equation (2) applied to storm deposits by Pepe et al. [22] requires input on the slope value of the granodiorite bench under attack by storm waves. Introduction of the 30° angle observed in the crude layering of the conglomerate generates negative wave height values. As shown in the original fieldwork by Peppe et al. [22], the inclination of platform rocks is very small (less than 2.5°) on which blocks are potentially subject to plucking by storm waves. In the case of previous work on andesite rocky shores from Isla San Luis Gonzaga in the Gulf of California [6] and basalt rocky shores from Gran Canaria in the Canary Islands [11], it is assumed that the angle of bed slope is only 1° at the pretransport location for joint bound blocks. Herein, the same assumption is made for the granodiorite on the eastern flank of Isla San Diego, which yields results from the application of Equation (2) based on Pepe et al. [22] that are similar to the wave heights obtained by application of Equation (1) following the work of Nott [20], as summarized in Table 1.

In this case, the 30° angle of repose observed in the upper part of the conglomerate is interpreted as the slope acquired by the superimposed deposit spilling over a rocky ridge in shallow water and not related to an earlier structural tilting of the island. The conglomerate layers may be interpreted as the result of overwash at the southern end of the island by storm waves. Today, the maximum elevation between Stations 2 and 6 on opposite shores is 8 m at the top of the conglomerate (Figure 1b). During the Late Pleistocene time (Marine Isotope Substage 5e), the bedrock spur at the island's south end was easily vulnerable to impact by exceptionally large waves. Although limited in scope, the available evidence for clast imbrication within the Isla San Diego conglomerate is consistent with the arrival of wind-driven waves from the northeast or east related to the shifting northward passage of a hurricane. In further consideration of the tilted layering in its upper part, a geometric solution for the true thickness of the deposit perpendicular to the dip angle amounts to about 4 m with the reactivation surface located roughly in the middle. The implication of the reactivation surface is that a second phase took place in storm energy, or that a separate storm event occurred sometime after the passage of the first event.

### 5.2. Significance of the Fossil Fauna from Isla San Diego

A strong latitudinal bias is reported in a comprehensive survey of the stony corals now living in the Gulf of California with 38 species occurring in the southern (or lower) part of the gulf, out of which only 16 have an extended range into the northern (or upper) part of the gulf [27]. In contrast, only three stony corals are limited to the upper Gulf of California with no representatives in the lower gulf. The disparity in today's geographic distribution is strongly related to the north–south temperature gradient. Pleistocene coral reefs are well preserved at several locations throughout the Gulf of California as far north as latitude 28° (Figure 1). The species *Porites panamensis* is ubiquitous in Upper Pleistocene coral reefs throughout the peninsular shores and gulf islands of Baja California Sur [24], and this coral species is the primary component. For example, *P. panamensis* is the principal component of Upper Pleistocene reefs on Isla Cerralvo [23], located 120 km farther to the southeast from Isla San Diego in the lower Gulf of California (Figure 1a). The north–south distance between the two islands is a full degree of latitude. In addition to *P. panamensis* (Figure 7a), the occurrence of *Pocillopora elegans* (Figure 7b), together with *Pavona gigantea* (Figure 7c), suggests all three were part of a former reef community that no longer persists around the island.

### 5.3. Maine Circulation and Recent Hurricanes in the Region of Isla San Diego

On an annual basis from November to May, strong winds capable of generating large-scale wave trains travel from north to south over the entire length of the Gulf of California, but lighter winds typically blow in the opposite direction under the influence of a semi-monsoonal pattern of atmospheric circulation during the spring and summer times [28]. Winds out of the north set up long-shore currents that may be responsible for delivering granodiorite gravel and perhaps even some cobbles from the flanks of Isla San Diego to the linked underwater ridge. In contrast, the lighter southerly winds probably are cable of shifting only pebbles.

The threat of hurricanes affecting the lower Gulf of California is reviewed in prior studies on storm deposits at Ensenada Almeja north of Loreto [4], Arroyo Blanco on Isla del Carmen west of Loreto [3] Puerto Escondido south of Loreto [5], and especially on Isla Cerralvo east of La Paz [23] (see Figure 1a for geographic relationships). Between September 1996 and September 2019, six named hurricanes (Fausto, Marty, Ignacio, John, Odile, and Lorena) entered the Gulf of California after originating farther south off the mainland coast of Mexico around Acapulco (Figure 1a). Storm rotation is counter-clockwise and as winds intensify during the northward passage, the greatest impact is expected to come from wind-driven waves pushing east to west out of the storm's northeast quarter. Eye-witness accounts of coastal wave surge during hurricane events in the Gulf of California are not common, but the published account of the Holocene storm beds at Ensenada Almeja includes a video clip of the 9 m storm surge against the rocky shore on nearby Ensenada San Basilio during Hurricane Odile [4]. Storm waves pushed by winds streaming from east to west during the passage of the same event farther south at Isla San Diego can be expected to have caused a surge that topped over the low-lying southern end of the island.

The 2015 hurricane season was unusually active [29], and Hurricane Patricia was a Category 5 event with wind speeds of 346 km/h that took an unexpected easterly turn north of Acapulco to strike the village of Cuixmala below the opening to the Gulf of California (Figure 1a). It remains one of the largest storms recorded in the eastern Pacific basin and the strongest yet to strike western Mexico. With a diameter of 2400 km, the storm's outer wind bands already swept across the tip of the Baja California Peninsula before the center veered eastward. Had the storm continued onward in its expected track to the northeast and gained strength from warmer waters in the Gulf of California, major damage from storm surge was certain to have occurred. Isla San Diego is one of those remote spots in the lower Gulf of California that would have experienced the full impact of wave shock against its east-facing shores and likely overwash of eroded materials across the bedrock at the island's southern tip.

Research based on the tagging of larger boulders from the conglomerate on Isla San Diego and its related submarine ridge would be instrumental in documenting the role of future hurricanes as a source of ongoing erosion. As part of a potentially larger project, the same monitoring program could be undertaken for the Pleistocene and Holocene storm deposits at Isla del Carmen [3], Ensenada Almeja [4], Puerto Escondido [5], and Isla San Luis Gonzaga [6]. The relevance of such a program is underscored by the record-breaking early start to the eastern Pacific hurricane season with the formation of tropical storm Andres located 960 km south of the tip of the Baja California Peninsula on 9 May 2021 [30]. The head start of the hurricane season in this part of the world portends the consequences of increased global warming.

*5.4. Comparison with Storm Deposits Elsewhere in the Gulf of California*

Variations in density among rock types studied so far from Pleistocene and Holocene boulder beds around the Gulf of California range from 1.86 g/cm$^3$ for limestone, 2.16 g/cm$^3$ for the banded rhyolite, and as much as 2.55 g/cm$^3$ for andesite based on different localities [3–6]. Estimates for the average wave height based on the average weight of sampled boulders taking rock density into account vary between 4.3 m to 5.7 m from locality to locality. However, when the largest boulders from each of four localities previously studied are taken into account, wave heights necessary for dislodgment from joint-bound basement rocks fall between 9.8 and more than 13 m. By comparison, the granodiorite samples from Isla San Diego register average wave heights rather smaller, between 2.9 and 3.1 m, based on the average of average estimates from six sample stations. However, the largest single granodiorite boulder measured on the island is estimated to weigh nearly two metric tons and to have required a wave height between 10.6 and 11.4 m to achieve dislodgement (Table 1). In this regard, the maximum wave shock that affected Isla San Diego during Pleistocene time is not out of the ordinary for the localities studied elsewhere in the Gulf of California. In terms of future hurricane events certain to strike the lower Gulf of California, an interesting prospect would be to tag some of the largest granodiorite boulders on the shallow ridge extending southwest of the island to see if movement occurs following the next major storm. Likewise, it could be interesting to tag some of the cobbles at the top of the crudely layered conglomerate beds currently at an elevation of 8 m to see if the next major storm creates waves capable of washing over that part of the island.

*5.5. Comparison with Storm Deposits Elsewhere in the North Atlantic Ocean*

Studies on boulder beds from the Pleistocene of Gran Canaria in the Canary Islands [11] and the Pleistocene and Holocene of Santa Maria Island in the Azores [10], as well as the Holocene of north Norway [12], follow the same format as those in Mexico's the Gulf of California using the triangular plots after Sneed and Folk [16] to appraise boulder shape and the same equations after Nott [20] and Peppe et al. [22] to estimate wave heights. Basalt boulders from El Copnfital Beach on Gran Canaria register a rock density of 2.84 g/cm$^3$, whereas those from Santa Maria Island in the Azores were treated as having an even higher rock density of 3.0 g/cm$^3$. Holocene beach cobbles and boulders on Leka Island in the subarctic of Norway were assigned an even higher rock density of 3.32 g/cm$^3$ associated with low-grade chromite ore [12]. All these rock densities from island localities in the North Atlantic Ocean surpass those for limestone, banded rhyolite, and andesite found at localities in the Gulf of California. Other things being equal in terms of volume, it requires a larger wave to extract a block of denser material such as basalt from a rocky shoreline than for material much less dense such as limestone. Based on the average weight and rock density of all basalt clasts measured from Gran Canaria, wave heights were 4.5 m, whereas based only on the largest boulders from each sample station, the maximum wave height was more than twice that value at 11 m. The results for basalt clasts from Santa Maria Island in the Azores were significantly less in both categories with an average weight from all sample stations, amounting to 2.6 m, whereas the maximum wave height based on the single largest boulder from each sample station amounted to 5.1 m. For storm deposits from Norway's Leka Island, the average results for cobbles and boulders derived from chromite ore from three sample stations estimated wave heights between 3.6 and 4.3 m. However, the average wave heights based on the single largest boulders from those stations yielded values for wave heights between 5.1 and 6.7 m. A comparison with the North Atlantic data on this basis puts the results from Isla San Diego closest in the range of values obtained from the Canary Islands. Essentially, storms of hurricane strength reaching the Gulf of California are no less severe in terms of their erosional effect than those in the northeastern Atlantic Ocean.

## 6. Conclusions

Study of the cobble–boulder deposits from Isla San Diego in Mexico's lower Gulf of California offers the following insights based on mathematical equations for estimation of Late Pleistocene wave heights from major storms in the same region:

- Consolidated cobbles and boulders studied from six sample stations with Upper Pleistocene conglomerate exhibit evidence of high-energy erosion from granodiorite exposed along the rocky shoreline of Isla San Diego in Mexico's lower Gulf of California. Evidence of clast imbrication indicates that a major storm had an impact with wave surge against the island's eastern shore;
- The average estimated volume at 66,536 cm$^3$ and the average weight of individual granodiorite cobbles and boulders at 100 kg from a total of 200 samples suggest that wave heights of 3 m are responsible for their derivation from the adjacent and joint-bound body of parent rock. However, the largest igneous boulder from among all six sample sites is estimated to weigh two metric tons and may have been moved by a wave of extraordinary height around 10 m. Alternately, smaller waves may have gradually loosened this block from its parent body until the force of gravity entrained it within the conglomerate;

- Compared to other localities in the Gulf of California where sea cliffs composed of igneous rocks such as andesite or banded rhyolite shed Holocene boulders, the granodiorite from Isla San Diego includes a larger fraction of elongated boulders that were more bar-like in form when originally loosened from the parent sea cliffs;
- At a higher rock density than local limestone or rhyolite at 1.86 g/cm$^3$ and 2.16 g/cm$^3$, respectively, granodiorite at 2.52 g/cm$^3$ required more wave energy for shore erosion. However, the difference between the measured rock density of local andesite only slightly exceeds that for local granodiorite and made little difference;
- Fossils recovered from granodiorite conglomerate on Isla San Diego expand the range distribution of reef-dwelling corals such as *Pocillopoora* and *Povona* farther northward than previously known. Otherwise, fossil representatives among the mollusks are typical of faunas more widely attributed Marine Isotope Substage 5e throughout the Gulf of California.

**Author Contributions:** M.E.J. was responsible for conceptualization and methodology of the project as a contribution to the Special Issue in the *Journal of Marine Sciences and Engineering* devoted to "Evaluation of Boulder Deposits Linked to Late Neogene Hurricane Events." G.C. was responsible for validation during field work at the site in May 2020 and March 2021. R.G.-F. was responsible for software applications and formal analysis. J.L.-V. was responsible for compilation of resources. All authors have read and agreed to the published version of the manuscript.

**Funding:** This project received no outside funding.

**Institutional Review Board Statement:** Not applicable.

**Informed Consent Statement:** Not applicable.

**Data Availability Statement:** All data are available in the published manuscript.

**Acknowledgments:** M.E.J. is grateful to Jay Racela (Environmental Lab, Williams College) for help with the experimental calculation of density for the granodiorite sample from the lower Gulf of California. Three readers contributed peer reviews with useful comments that led to the improvement of this contribution.

**Conflicts of Interest:** The authors declare no conflict of interest.

## Appendix A

**Table A1.** Quantification of cobble and boulder sizes, volume, and estimated weight from Station 1 near the south end of Isla San Diego. The density of granite at 2.52 g/cm$^3$ is applied uniformly in order to calculate wave height for each boulder on the basis of competing equations. Abbreviation: EWH = estimated wave height. Coordinates 25.11.7009 N and 110.42.0833 W.

| Sample | Long Axis (cm) | Intermediate Axis (cm) | Short Axis (cm) | Volume (cm$^3$) | Adjust to 80% | Weight (kg) | EWH Nott [20] (m) | EWH Pepe et al. [22] (m) |
|---|---|---|---|---|---|---|---|---|
| 1 | 118 | 75 | 31 | 274,350 | 219,480 | 553 | 9.7 | 5.8 |
| 2 | 24 | 20 | 7 | 3360 | 2688 | 6.8 | 2.0 | 1.3 |
| 3 | 79 | 53 | 32 | 133,984 | 107,187 | 270 | 6.5 | 6.0 |
| 4 | 39 | 34 | 21 | 27,846 | 22,277 | 56 | 3.2 | 3.9 |
| 5 | 40 | 30 | 28 | 33,600 | 26,880 | 68 | 3.3 | 5.2 |
| 6 | 68 | 53 | 23 | 82,892 | 66,314 | 167 | 5.6 | 4.3 |
| 7 | 68 | 43 | 16 | 46,784 | 37,427 | 94 | 5.6 | 3.0 |
| 8 | 120 | 110 | 70 | 924,000 | 739,200 | 1863 | 9.9 | 13.1 |
| 9 | 65 | 43 | 29 | 81,055 | 64,844 | 163 | 5.4 | 5.4 |
| 10 | 43.5 | 32 | 18 | 25,056 | 20,045 | 51 | 3.6 | 3.4 |
| 11 | 32 | 21 | 12 | 8064 | 6451 | 16 | 2.6 | 2.2 |
| 12 | 29 | 26 | 14.5 | 10,933 | 8746 | 22 | 2.4 | 2.7 |
| 13 | 68 | 55 | 23 | 86,020 | 68,816 | 173 | 5.6 | 4.3 |
| 14 | 41 | 34 | 14 | 19,516 | 15,613 | 39 | 3.4 | 2.6 |
| 15 | 64 | 33 | 15 | 31,680 | 25,344 | 64 | 5.3 | 2.8 |

**Table A1.** *Cont.*

| Sample | Long Axis (cm) | Intermediate Axis (cm) | Short Axis (cm) | Volume (cm$^3$) | Adjust to 80% | Weight (kg) | EWH Nott [20] (m) | EWH Pepe et al. [22] (m) |
|---|---|---|---|---|---|---|---|---|
| 16 | 90 | 67 | 26 | 156,780 | 125,424 | 316 | 7.4 | 4.9 |
| 17 | 17 | 14 | 8 | 1904 | 1523 | 3.8 | 1.4 | 1.5 |
| 18 | 31 | 26 | 11 | 8866 | 7093 | 17.9 | 2.6 | 2.1 |
| 19 | 15 | 11 | 2.5 | 413 | 330 | 0.8 | 1.2 | 0.5 |
| 20 | 12 | 12 | 4.5 | 648 | 518 | 1.3 | 1.0 | 0.8 |
| 21 | 10.5 | 10.5 | 3 | 331 | 265 | 0.7 | 0.9 | 0.6 |
| 22 | 128 | 125 | 61 | 976,000 | 780,800 | 1968 | 10.6 | 11.4 |
| 23 | 72 | 68 | 67 | 328,032 | 262,426 | 661 | 5.9 | 12.5 |
| 24 | 16 | 14 | 3 | 672 | 538 | 1.4 | 1.3 | 0.6 |
| 25 | 50 | 48 | 27 | 64,800 | 51,840 | 131 | 4.1 | 5.0 |
| 26 | 18 | 16 | 7 | 2016 | 1613 | 4.1 | 1.5 | 1.3 |
| 27 | 13 | 12 | 4 | 624 | 499 | 1.3 | 1.1 | 0.7 |
| 28 | 30 | 16.5 | 14 | 6930 | 5544 | 14 | 2.5 | 2.6 |
| 29 | 20 | 12 | 6 | 1440 | 1152 | 2.9 | 1.7 | 1.1 |
| 30 | 18 | 16 | 6.5 | 1872 | 1498 | 3.8 | 1.5 | 1.2 |
| Average | 47.96 | 37.66 | 20.13 | 111,349 | 89,079 | 224.0 | 3.2 | 3.0 |

**Table A2.** Quantification of cobble and boulder sizes and volume with estimated weight from Station 2 near the south end of Isla San Diego. The density of granite at 2.52 g/cm$^3$ is applied uniformly in order to calculate wave height for each boulder on the basis of competing equations. Abbreviation: EWH = estimated wave height. Coordinates 25.11.6713 N and 110.42.1023 W.

| Sample | Long Axis (cm) | Intermediate Axis (cm) | Short Axis (cm) | Volume (cm$^3$) | Adjust to 80% | Weight (kg) | EWH Nott [20] (m) | EWH Pepe et al. [22] (m) |
|---|---|---|---|---|---|---|---|---|
| 1 | 23.5 | 12 | 12 | 3384 | 2707 | 6.8 | 1.9 | 2.2 |
| 2 | 15 | 11 | 9 | 1485 | 1188 | 3 | 1.2 | 1.7 |
| 3 | 31 | 29.5 | 11 | 10,060 | 8048 | 20 | 2.6 | 2.1 |
| 4 | 19 | 13 | 7 | 1729 | 1383 | 3.5 | 1.6 | 1.3 |
| 5 | 16 | 9.5 | 4.5 | 684 | 547 | 1.4 | 1.3 | 0.8 |
| 6 | 13.5 | 9 | 7 | 851 | 680 | 1.7 | 1.1 | 1.3 |
| 7 | 16.5 | 14.5 | 8 | 1798 | 1438 | 3.6 | 1.4 | 1.5 |
| 8 | 18 | 14 | 8 | 2016 | 1613 | 4 | 1.5 | 1.5 |
| 9 | 26 | 21 | 9 | 4914 | 3913 | 10 | 2.1 | 1.7 |
| 10 | 21 | 10 | 8 | 1680 | 1344 | 3.4 | 1.7 | 1.5 |
| 11 | 34 | 27 | 15.5 | 14,229 | 11,383 | 28.7 | 2.8 | 2.9 |
| 12 | 43 | 32 | 17 | 23,392 | 18,714 | 476 | 3.6 | 3.2 |
| 13 | 17 | 10.5 | 8.5 | 1517 | 1214 | 3.1 | 1.4 | 1.6 |
| 14 | 20.5 | 13 | 9 | 2399 | 1919 | 4.8 | 1.7 | 1.7 |
| 15 | 16 | 8 | 3 | 384 | 307 | 0.8 | 1.3 | 0.6 |
| 16 | 34 | 33 | 30 | 33,660 | 26.928 | 67.9 | 2.8 | 5.6 |
| 17 | 33 | 31 | 16 | 16,368 | 13,094 | 33 | 2.7 | 3.0 |
| 18 | 29 | 24 | 9 | 6264 | 5011 | 12.6 | 2.4 | 1.7 |
| 19 | 40 | 36 | 36 | 51,840 | 41,472 | 105 | 3.3 | 6.7 |
| 20 | 62 | 45 | 38 | 106,020 | 84,816 | 214 | 5.1 | 7.1 |
| 21 | 59 | 28 | 24 | 39,648 | 31,714 | 80 | 4.9 | 4.5 |
| 22 | 16.5 | 13 | 4 | 858 | 686 | 1.7 | 1.4 | 0.7 |
| 23 | 15.5 | 13 | 5.5 | 1108 | 887 | 2.2 | 1.3 | 1.0 |
| 24 | 12 | 11 | 5 | 660 | 528 | 1.3 | 1.0 | 0.9 |
| 25 | 26.5 | 19 | 15 | 7553 | 6042 | 15 | 2.2 | 2.8 |
| 26 | 9 | 8 | 3 | 216 | 173 | 0.4 | 0.7 | 0.6 |
| 27 | 14 | 12.5 | 7 | 1225 | 980 | 2.5 | 1.2 | 1.3 |
| 28 | 10 | 9 | 7 | 630 | 504 | 1.3 | 0.8 | 1.3 |
| 29 | 16 | 9.5 | 4 | 608 | 486 | 1.2 | 1.3 | 0.7 |

**Table A2.** *Cont.*

| Sample | Long Axis (cm) | Intermediate Axis (cm) | Short Axis (cm) | Volume (cm$^3$) | Adjust to 80% | Weight (kg) | EWH Nott [20] (m) | EWH Pepe et al. [22] (m) |
|---|---|---|---|---|---|---|---|---|
| 30 | 50 | 39 | 12.5 | 24,375 | 19,500 | 49 | 4.1 | 2.3 |
| 31 | 31 | 19.5 | 10 | 6045 | 4836 | 12 | 2.6 | 1.9 |
| 32 | 28 | 15 | 6 | 2520 | 2016 | 5.1 | 2.3 | 1.1 |
| 33 | 8.5 | 5.5 | 4 | 182 | 150 | 0.4 | 0.7 | 0.7 |
| 34 | 12.5 | 9 | 7 | 788 | 630 | 1.6 | 1.0 | 1.3 |
| 35 | 23 | 19 | 5 | 2185 | 1748 | 4.4 | 1.6 | 0.9 |
| Average | 24.5 | 18.0 | 11.0 | 10,665 | 7763 | 33.8 | 1.9 | 1.9 |

**Table A3.** Quantification of cobble and boulder sizes, volume, and estimated weight from Station 3 at the south end of Isla San Diego. The density of granite at 2.52 g/cm$^3$ is applied uniformly in order to calculate wave height for each boulder on the basis of competing equations. Abbreviation: EWH = estimated wave height. Coordinates 25.11.6426 N and 110.42. 2518 W.

| Sample | Long Axis (cm) | Intermediate Axis (cm) | Short Axis (cm) | Volume (cm$^3$) | Adjust to 80% | Weight (kg) | EWH Nott [20] (m) | EWH Pepe et al. [22] (m) |
|---|---|---|---|---|---|---|---|---|
| 1 | 68 | 61 | 30 | 124,440 | 99,552 | 251 | 5.6 | 5.6 |
| 2 | 19 | 12.5 | 6 | 1425 | 1140 | 2.9 | 1.6 | 1.1 |
| 3 | 20 | 16 | 6.5 | 2080 | 1664 | 4.2 | 1.7 | 1.2 |
| 4 | 52 | 36 | 26 | 48,672 | 38,938 | 98 | 4.3 | 4.9 |
| 5 | 18.5 | 16 | 6.5 | 1924 | 1539 | 3.9 | 1.5 | 1.2 |
| 6 | 63 | 54 | 34 | 115,668 | 92,534 | 233 | 5.2 | 6.3 |
| 7 | 52 | 36 | 26 | 48,672 | 38,938 | 98 | 4.3 | 4.9 |
| 8 | 61 | 25 | 22 | 33,550 | 26,840 | 68 | 5.0 | 4.1 |
| 9 | 26.5 | 20 | 13 | 6890 | 5512 | 14 | 2.2 | 2.4 |
| 10 | 32.5 | 16.5 | 12 | 6435 | 5148 | 13 | 2.7 | 2.2 |
| 11 | 27 | 20 | 10 | 5400 | 4320 | 10.9 | 2.2 | 1.9 |
| 12 | 101 | 38 | 37.5 | 143,925 | 115,140 | 290 | 8.3 | 7.0 |
| 13 | 35 | 33 | 14.5 | 16,748 | 13,398 | 34 | 2.9 | 2.7 |
| 14 | 62 | 45 | 43.5 | 121,365 | 97,092 | 245 | 5.1 | 8.1 |
| 15 | 51 | 38 | 23 | 44,574 | 35,659 | 90 | 4.2 | 4.3 |
| 16 | 18.5 | 13.5 | 10.5 | 2622 | 2098 | 5.3 | 1.5 | 2.0 |
| 17 | 15 | 12 | 10 | 1800 | 1440 | 3.6 | 1.2 | 1.9 |
| 18 | 126 | 59.5 | 50 | 374,850 | 299,850 | 756 | 10.4 | 9.3 |
| 19 | 20.5 | 19 | 13 | 5064 | 4051 | 10 | 1.7 | 2.4 |
| 20 | 71 | 54 | 28 | 102,352 | 85,882 | 216 | 5.9 | 5.2 |
| 21 | 49 | 22 | 17 | 18,326 | 14,661 | 37 | 4.0 | 3.2 |
| 22 | 17.5 | 13 | 6 | 1365 | 1092 | 2.8 | 1.4 | 1.1 |
| 23 | 28 | 27 | 17 | 12,852 | 10,282 | 26 | 2.3 | 3.2 |
| 24 | 10.5 | 8 | 5 | 420 | 336 | 0.8 | 0.9 | 0.9 |
| 25 | 21 | 18 | 12.5 | 4725 | 3780 | 9.5 | 1.7 | 2.3 |
| 26 | 12.5 | 7.5 | 5.5 | 516 | 413 | 1.0 | 1.0 | 1.0 |
| 27 | 13 | 11.5 | 8 | 1196 | 957 | 2.4 | 1.1 | 1.5 |
| 28 | 11.5 | 7 | 4 | 322 | 258 | 0.7 | 1.0 | 0.7 |
| 29 | 16.5 | 14 | 8 | 1848 | 1478 | 3.7 | 1.4 | 1.5 |
| 30 | 11 | 6 | 3.5 | 231 | 185 | 0.5 | 0.9 | 0.7 |
| 31 | 56 | 29 | 23 | 37,352 | 24,882 | 75 | 4.6 | 4.3 |
| 32 | 19 | 13 | 4.5 | 9182 | 7346 | 19 | 1.6 | 0.8 |
| 33 | 14 | 13 | 6.5 | 1183 | 946 | 2.4 | 1.2 | 1.2 |
| 34 | 15 | 12 | 8 | 1440 | 1152 | 2.9 | 1.2 | 1.5 |
| 35 | 28 | 25.5 | 13 | 9282 | 7426 | 19 | 2.3 | 2.4 |
| Average | 36.0 | 24.32 | 16.0 | 37,391 | 29,884 | 75.7 | 2.8 | 2.8 |

**Table A4.** Quantification of cobble and boulder sizes, volume, and estimated weight from Station 4 at the south end of Isla San Diego. The density of granite at 2.52 g/cm$^3$ is applied uniformly in order to calculate wave height for each boulder on the basis of competing equations. Abbreviation: EWH = estimated wave height. Coordinates 25.11.6463 N and 110.42. 2586 W.

| Sample | Long Axis (cm) | Intermediate Axis (cm) | Short Axis (cm) | Volume (cm$^3$) | Adjust to 80% | Weight (kg) | EWH Nott [20] (m) | EWH Pepe et al. [22] (m) |
|---|---|---|---|---|---|---|---|---|
| 1 | 66 | 49 | 30 | 97,020 | 77,616 | 196 | 5.5 | 5.6 |
| 2 | 46 | 36 | 14 | 23,184 | 18,547 | 47 | 3.8 | 2.6 |
| 3 | 41 | 26.5 | 14 | 15,211 | 12,169 | 31 | 3.4 | 2.6 |
| 4 | 30 | 21 | 14.5 | 9135 | 7308 | 18 | 2.5 | 2.7 |
| 5 | 41.5 | 36 | 11.5 | 17,181 | 13,745 | 35 | 3.4 | 2.1 |
| 6 | 70.5 | 52 | 29 | 106,314 | 85,051 | 214 | 5.8 | 5.4 |
| 7 | 57 | 21 | 20 | 23,940 | 19,152 | 48 | 4.7 | 3.7 |
| 8 | 33.5 | 28 | 13.5 | 12,663 | 10,130 | 26 | 2.8 | 2.5 |
| 9 | 30 | 29.5 | 14 | 12,390 | 9912 | 25 | 2.5 | 2.6 |
| 10 | 50 | 36.5 | 14 | 25,550 | 20,440 | 52 | 4.1 | 2.6 |
| 11 | 60.5 | 33 | 19.5 | 38,932 | 31,145 | 78 | 5.0 | 3.6 |
| 12 | 52 | 37 | 22 | 42,328 | 33,862 | 85 | 4.3 | 4.1 |
| 13 | 15 | 12.5 | 8.5 | 1594 | 1275 | 3.2 | 1.2 | 1.6 |
| 14 | 113 | 55 | 53 | 329,395 | 263,516 | 664 | 9.3 | 9.9 |
| 15 | 55 | 32 | 32 | 56,320 | 45,056 | 114 | 4.5 | 6.0 |
| 16 | 34 | 28 | 9 | 8568 | 6854 | 17 | 2.8 | 1.7 |
| 17 | 43 | 36.5 | 23 | 36,099 | 28,879 | 73 | 3.6 | 4.3 |
| 18 | 19 | 13.5 | 7 | 1796 | 1436 | 3.6 | 1.6 | 1.3 |
| 19 | 63 | 49 | 35 | 108,045 | 86,436 | 218 | 5.2 | 6.5 |
| 20 | 16 | 12.5 | 9.5 | 1900 | 1520 | 3.8 | 1.3 | 1.8 |
| 21 | 91.5 | 74 | 17.5 | 118,493 | 94,794 | 239 | 7.6 | 3.3 |
| 22 | 65 | 42.5 | 19 | 52,488 | 41,990 | 106 | 5.4 | 3.5 |
| 23 | 89 | 35 | 32.5 | 101,238 | 80,990 | 204 | 7.4 | 6.1 |
| 24 | 55 | 42 | 22.5 | 51,975 | 41,580 | 105 | 4.5 | 4.2 |
| 25 | 31 | 15.5 | 11.5 | 5526 | 4421 | 11 | 2.6 | 2.1 |
| 26 | 40.5 | 15.5 | 9.5 | 5,964 | 4,771 | 12 | 3.3 | 1.8 |
| 27 | 64 | 44 | 35 | 98,560 | 78,848 | 199 | 5.3 | 6.5 |
| 28 | 28 | 19 | 18.5 | 9842 | 7874 | 20 | 2.3 | 3.5 |
| 29 | 56.5 | 35 | 13 | 25,708 | 20,566 | 52 | 4.7 | 2.4 |
| 30 | 13 | 9 | 6.5 | 761 | 608 | 1.5 | 1.1 | 1.2 |
| 31 | 16 | 10 | 7 | 1120 | 896 | 2.3 | 1.3 | 1.3 |
| 32 | 19.5 | 12 | 9.5 | 2223 | 1778 | 4.5 | 1.6 | 1.8 |
| 33 | 98 | 47 | 26 | 119,756 | 95,805 | 241 | 8.1 | 4.9 |
| 34 | 19.5 | 17 | 11 | 3647 | 2917 | 7.4 | 1.4 | 2.1 |
| 35 | 57.5 | 56 | 44 | 141,680 | 113,344 | 286 | 4.8 | 8.2 |
| Average | 48.0 | 32.0 | 19.31 | 48,758 | 39,007 | 98.0 | 3.7 | 3.4 |

**Table A5.** Quantification of cobble and boulder sizes, volume, and estimated weight from Station 5 at the south end of Isla San Diego. The density of granite at 2.52 g/cm$^3$ is applied uniformly in order to calculate wave height for each boulder on the basis of competing equations. Abbreviation: EWH = estimated wave height. Coordinates 25.11.6494 N and 110.42.2551 W.

| Sample | Long Axis (cm) | Intermediate Axis (cm) | Short Axis (cm) | Volume (cm$^3$) | Adjust to 80% | Weight (kg) | EWH Nott [20] (m) | EWH Pepe et al. [22] (m) |
|---|---|---|---|---|---|---|---|---|
| 1 | 48.5 | 40.5 | 20 | 39,285 | 31,428 | 79.8 | 4.0 | 3.7 |
| 2 | 64 | 33 | 23.5 | 49,632 | 39,706 | 100 | 5.3 | 4.4 |
| 3 | 69 | 31 | 28.5 | 60,962 | 48,769 | 123 | 5.7 | 5.3 |
| 4 | 36 | 35 | 20.5 | 25,830 | 20,664 | 52 | 3.0 | 3.8 |
| 5 | 73 | 43 | 25 | 78,475 | 62,780 | 158 | 6.0 | 4.7 |
| 6 | 25 | 22.5 | 13 | 7313 | 5850 | 15 | 2.1 | 2.4 |
| 7 | 33 | 23 | 12 | 9108 | 7286 | 18 | 2.7 | 2.2 |

Table A5. *Cont.*

| Sample | Long Axis (cm) | Intermediate Axis (cm) | Short Axis (cm) | Volume (cm$^3$) | Adjust to 80% | Weight (kg) | EWH Nott [20] (m) | EWH Pepe et al. [22] (m) |
|---|---|---|---|---|---|---|---|---|
| 8 | 36 | 36 | 11.5 | 14,904 | 4923 | 30 | 3.0 | 2.1 |
| 9 | 13 | 7.5 | 7 | 683 | 546 | 1.4 | 1.1 | 1.3 |
| 10 | 21 | 7.5 | 6.5 | 1024 | 819 | 2 | 1.7 | 1.2 |
| 11 | 14 | 13 | 8.5 | 1547 | 1238 | 3.1 | 1.2 | 1.6 |
| 12 | 49.5 | 43 | 19.5 | 41,506 | 33,205 | 84 | 4.1 | 3.6 |
| 13 | 19.5 | 17 | 10.5 | 3481 | 2785 | 7 | 1.6 | 2.0 |
| 14 | 49.5 | 29 | 22 | 31,581 | 25,265 | 64 | 4.1 | 4.1 |
| 15 | 31 | 30 | 5 | 4650 | 3720 | 9.4 | 2.6 | 0.9 |
| 16 | 11 | 9 | 6.5 | 644 | 515 | 1.3 | 0.9 | 1.2 |
| 17 | 9.5 | 8 | 7 | 532 | 426 | 1.1 | 0.8 | 1.3 |
| 18 | 31.5 | 30 | 16 | 15,120 | 12,096 | 30 | 2.6 | 3.0 |
| 19 | 42 | 38 | 14.5 | 23,142 | 18,514 | 47 | 3.5 | 2.7 |
| 20 | 45 | 45 | 22 | 44,550 | 35,640 | 90 | 3.7 | 4.1 |
| 21 | 63 | 40 | 26 | 65,520 | 52,416 | 132 | 5.2 | 4.9 |
| 22 | 67 | 36 | 22 | 53,064 | 42,45 | 107 | 5.5 | 4.1 |
| 23 | 63 | 53 | 29.5 | 98,501 | 78,800 | 199 | 5.2 | 5.5 |
| 24 | 40 | 30 | 16.5 | 19,800 | 15,840 | 40 | 3.3 | 3.1 |
| 25 | 28 | 19.5 | 14 | 7644 | 6115 | 15 | 2.3 | 2.6 |
| 26 | 84.5 | 58.5 | 28 | 138,411 | 110,729 | 279 | 7.0 | 5.2 |
| 27 | 63 | 39 | 28 | 63,504 | 50,803 | 128 | 5.2 | 5.2 |
| 28 | 78.5 | 62 | 25 | 121,675 | 97,340 | 245 | 6.5 | 4.7 |
| 29 | 65 | 46 | 38 | 113,620 | 90,896 | 229 | 5.4 | 7.1 |
| 30 | 29 | 28 | 9.5 | 7714 | 6171 | 16 | 2.4 | 1.8 |
| 31 | 100.5 | 45.5 | 43 | 196,628 | 157,303 | 396 | 8.3 | 8.0 |
| 32 | 37 | 32 | 16.5 | 19,536 | 15,629 | 39 | 3.1 | 3.1 |
| 33 | 66 | 29.5 | 28 | 54,516 | 43,613 | 40 | 5.5 | 5.2 |
| 34 | 58 | 45.5 | 24 | 63,336 | 50,669 | 128 | 4.8 | 4.5 |
| 35 | 32 | 23 | 8.5 | 6256 | 5005 | 13 | 2.6 | 1.6 |
| Average | 45.58 | 32.24 | 18.73 | 43,391 | 33,456 | 83.4 | 3.6 | 3.3 |

Table A6. Quantification of cobble and boulder sizes, volume, and estimated weight from Station 6 at the south end of Isla San Diego. The density of granite at 2.52 g/cm$^3$ is applied uniformly in order to calculate wave height for each boulder on the basis of competing equations. Abbreviation: EWH = estimated wave height. Coordinates 25.11.6802 N and 110.42.2417 W.

| Sample | Long Axis (cm) | Intermediate Axis (cm) | Short Axis (cm) | Volume (cm$^3$) | Adjust to 80% | Weight (kg) | EWH Nott [20] (m) | EWH Pepe et al. [22] (m) |
|---|---|---|---|---|---|---|---|---|
| 1 | 75 | 58 | 27 | 117,450 | 93,960 | 238 | 6.2 | 5.0 |
| 2 | 69 | 33 | 14 | 31,878 | 25,502 | 64 | 5.7 | 2.6 |
| 3 | 50 | 45 | 35 | 78,750 | 63,000 | 159 | 4.1 | 6.5 |
| 4 | 64.5 | 39 | 35.5 | 89,300 | 71,440 | 180 | 5.3 | 6.6 |
| 5 | 30 | 24 | 13.5 | 9720 | 7776 | 20 | 2.5 | 2.5 |
| 6 | 25.5 | 14 | 14 | 4998 | 3998 | 10 | 2.1 | 2.6 |
| 7 | 60 | 38 | 26 | 59,280 | 47,424 | 120 | 5.0 | 4.9 |
| 8 | 67 | 42 | 19 | 53,466 | 42,773 | 108 | 5.5 | 3.5 |
| 9 | 53.5 | 28 | 25 | 37,450 | 29,960 | 75 | 4.4 | 4.7 |
| 10 | 19.5 | 18 | 8 | 2808 | 2246 | 5.7 | 1.6 | 1.5 |
| 11 | 73 | 35 | 20.5 | 52,378 | 41,902 | 106 | 6.0 | 3.8 |
| 12 | 63 | 54 | 25 | 85,050 | 68,040 | 171 | 5.2 | 4.7 |
| 13 | 85 | 56.5 | 19.5 | 93,649 | 74,919 | 189 | 4.7 | 3.6 |
| 14 | 76 | 43.5 | 30 | 99,180 | 79,344 | 200 | 6.3 | 5.6 |
| 15 | 63 | 40.5 | 19 | 48,479 | 38,783 | 98 | 5.2 | 3.5 |
| 16 | 27 | 26.5 | 11.5 | 8228 | 6583 | 17 | 2.2 | 2.1 |
| 17 | 35.5 | 22 | 14 | 10,934 | 8747 | 22 | 2.9 | 2.6 |
| 18 | 49 | 41 | 15.5 | 31,140 | 24,912 | 63 | 4.0 | 2.9 |

**Table A6.** *Cont.*

| Sample | Long Axis (cm) | Intermediate Axis (cm) | Short Axis (cm) | Volume (cm$^3$) | Adjust to 80% | Weight (kg) | EWH Nott [20] (m) | EWH Pepe et al. [22] (m) |
|---|---|---|---|---|---|---|---|---|
| 19 | 20.5 | 12.5 | 12 | 3075 | 2460 | 6.2 | 1.7 | 2.2 |
| 20 | 64 | 53 | 24.5 | 83,104 | 66,483 | 168 | 5.3 | 4.6 |
| 21 | 20.5 | 15.5 | 8.5 | 2701 | 2161 | 5.4 | 1.7 | 1.6 |
| 22 | 11 | 10 | 6 | 660 | 528 | 1.3 | 0.9 | 1.1 |
| 23 | 33.5 | 22 | 20 | 14,740 | 11,792 | 30 | 2.8 | 3.7 |
| 24 | 57 | 41 | 36.5 | 85,301 | 68,240 | 172 | 4.7 | 6.8 |
| 25 | 43 | 38.5 | 19 | 31,455 | 25,164 | 63 | 3.6 | 3.5 |
| 26 | 22.5 | 22 | 6.5 | 3218 | 2574 | 6.5 | 1.9 | 1.2 |
| 27 | 53 | 41 | 12 | 26,076 | 20,861 | 53 | 4.4 | 2.2 |
| 28 | 38 | 22.5 | 16 | 13,680 | 10,944 | 28 | 3.1 | 3.0 |
| 29 | 41 | 19.5 | 12.5 | 9994 | 7995 | 20 | 3.4 | 2.3 |
| 30 | 59 | 49 | 26 | 75,166 | 60,133 | 152 | 4.9 | 4.9 |
| Average | 48.3 | 33.5 | 19.0 | 42,110 | 33,688 | 85.0 | 3.2 | 2.9 |

## References


1. Carreño, A.L.; Helenes, J. Geology and ages of the islands. In *Island Biogeography of the Sea of Cortés*; Case, T.J., Cody, M.L., Ezcurra, E., Eds.; Oxford University Press: Oxford, UK, 2002; pp. 14–40.
2. Backus, D.H.; Johnson, M.E.; Ledesma-Vazquez, J. Peninsular and island rocky shores in the Gulf of California. In *Atlas of Coastal Ecosystems in the Western Gulf of California*; Johnson, M.E., Ledesma-Vazquez, J., Eds.; University Arizona Press: Tucson, AZ, USA, 2009; pp. 11–27. ISBN 978-0-8165-2530-0.
3. Johnson, M.E.; Ledesma-Vázquez, J.; Guardado-Grance, R. Coastal geomorphology of a Holocene hurricane deposit on a Pleistocene marine terrace from Isla Carmen (Baja California Sur, Mexico). *J. Mar. Sci. Eng.* **2018**, *6*, 108. [CrossRef]
4. Johnson, M.E.; Guardado-France, R.; Johnson, E.M.; Ledesma-Vázquez, J. Geomorphology of a Holocene hurricane deposit eroded from rhyolite sea cliffs on Ensenada Almeja (Baja California Sur, Mexico). *J. Mar. Sci. Eng.* **2019**, *7*, 193. [CrossRef]
5. Johnson, M.E.; Johnson, E.M.; Guardado-France, R.; Ledesma-Vázquez, J. Holocene hurricane deposits eroded as coastal barriers from andesite sea cliffs at Puerto Escondido (Baja California Sur, Mexico). *J. Mar. Sci. Eng.* **2020**, *8*, 75. [CrossRef]
6. Guardada-France, R.; Johnson, M.E.; Ledesma-Vázquez, J.; Santa Rosa-del Rio, M.A.; Herrera-Gutiérrez, Á. Multiphase storm deposits eroded from andesite sea cliffs on Isla San Luis Gonzaga (Northern Gulf of California Mexico). *J. Mar. Sci. Eng.* **2020**, *8*, 525. [CrossRef]
7. Lumbier, M.M. Algunos datos sobre las islas Mexicanas para contribuir al studio de sur recursos naturalas. *Anales del Instituo Geoólgico de México* **1919**, *7*, 29.
8. Lorang, M.S. A wave-competence approach to distinguish between boulder and megaclast deposits due to storm waves versus tsunamis. *Mar. Geol.* **2011**, *283*, 90–97. [CrossRef]
9. May, S.M.; Engel, M.; Brill, D.; Squire, P.; Scheffers, A.; Kelletat, D. Coastal hazards from tropical cyclones and extratropical winter storms based on Holocene storm chronologies. In *Coastal World Heritage Sites*; Finkl, C.W., Ed.; Springer Nature: Berlin, Germany, 2012; Volume 6, pp. 557–585.
10. Kennedy, A.B.; Mori, N.; Zhang, Y.; Yasuda, T.; Chen, S.E.; Tajima, Y.; Pecor, W.; Toride, K. Observations and modeling of coastal boulder transport and loading during Super Typhoon Haiyan. *Coast. Eng. J.* **2016**, *58*, 164004-1–164004-25. [CrossRef]
11. Scheffers, A.; Kelletat, D. Megaboulder movement by superstorms: A geomorphological approach. *J. Coast. Res.* **2020**, *36*, 844–856. [CrossRef]
12. Ávila, S.P.; Johnson, M.E.; Rebelo, A.C.; Baptista, L.; Melo, C.S. Comparison of modern and Pleistocene (MIS 5e) coastal boulder deposits from Santa Maria Island (Azores Archipelago, NE Atlantic Ocean). *J. Mar. Sci. Eng.* **2020**, *9*, 138. [CrossRef]
13. Galindo, I.; Johnson, M.E.; Martín-González, E.; Romero, C.; Vegas, J.; Melo, C.S.; Ávila, S.P.; Sánchez, N. Late Pleistocene boulder slumps eroded from a basalt shoreline at El Confital Beach on Gran Canaria (Canary Islands, Spain). *J. Mar. Sci. Eng.* **2020**, *9*, 138. [CrossRef]
14. Johnson, M.E. Holocene boulder beach eroded from chromite and dunite sea cliffs at Støpet on Leka Island (Northern Norway). *J. Mar. Sci. Eng.* **2020**, *8*, 644. [CrossRef]
15. Ledesma-Vázquez, J.; Johnson, M.E.; Gonzalez-Yajimovich, O.; Santamaría-del-Angel, E. Gulf of California geography, geological origins, oceanography, and sedimentation patterns. In *Atlas of Coastal Ecosystems in the Western Gulf of California*; Johnson, M.E., Ledesma-Vazquez, J., Eds.; University Arizona Press: Tucson, AZ, USA, 2009; pp. 1–10. ISBN 978-0-8165-2530-0.
16. Gastil, R.G.; Kimbrough, D.L.; Kimbrough, J.M. The Sierra San Pedro Mártir zoned pluton, Baja California, Mexico. In *Peninsular Ranges Batholith, Baja California and Southern California: Geological Society of America Memoirs*; Morton, D.M., Miller, F.K., Eds.; Geological Society of America: Boulder, CO, USA, 2014; Volume 211, pp. 739–758.

17. Olvera, M.C.; Aceves, J.S.; Rendøn, C.; Valiente, C.; Acosta, M.L.L.; Rodríguez, B. La política ambiental Mexicana y la conservación del ambiente en Baja California Sur. *Gaceta Ecológica* **2004**, *70*, 45–56.
18. Wentworth, C.K. A scale of grade and class terms for clastic sediments. *J. Geol.* **1922**, *27*, 377–392. [CrossRef]
19. Sneed, E.D.; Folk, R.L. Pebbles in the lower Colorado River of Texas: A study in particle morphogenesis. *J. Geol.* **1958**, *66*, 114–150. [CrossRef]
20. Nott, J. Waves, coastal bolder deposits and the importance of pre-transport setting. *Earth Planet. Sci. Lett.* **2003**, *210*, 269–276. [CrossRef]
21. Nandasena, N.A.K.; Paris, R.; Tanaka, N. Reassessment of hydrodynamic equations: Minimum flow velocity to initiate boulder transport by high energy events (storms, tsunamis). *Mar. Geol.* **2011**, *281*, 70–84. [CrossRef]
22. Pepe, F.; Corradino, M.; Parrino, N.; Besio, G.; Presti, V.L.; Renda, P.; Calcagnile, L.; Quarta, G.; Sulli, A.; Antonioli, F. Boulder coastal deposits at Favignana Island rocky coast (Sicily, Italy): Litho-structural and hydrodynamic control. *Geomorphology* **2018**, *303*, 191–209. [CrossRef]
23. Tierney, P.W.; Johnson, M.E. Stabilization role of crustose coralline algae during Late Pleistocene reef development on Isla Cerralvo, Baja California Sur (Mexico). *J. Coast. Res.* **2012**, *28*, 244–254. [CrossRef]
24. López-Pérez, R.A. Fossil corals from the Gulf of California, México: Still a depauperate fauna but it bears more species than previously thought. *Proc. Calif. Acad. Sci.* **2008**, *59*, 503–519.
25. Harmon, R.S.; Mitterer, R.M.; Kriausakul, N.; Land, L.S.; Schwarcz, H.P.; Garrett, P.; Larson, G.J.; Vacher, H.L.; Rowe, M. U-series and amino acid racemization geochronology of Bermuda—Implications for eustatic sea level fluctuation over the past 250,000 years. *Palaeogeogr. Palaeoclimatol. Palaeoecol.* **1983**, *44*, 41–70. [CrossRef]
26. Rohling, E.J.; Grant, K.; Hemleben, C.H.; Siddall, M.; Hoogakker, B.A.A.; Bolshaw, M.; Kucera, M. High rates of sea-level rise during the last interglacial period. *Nat. Geosci.* **2008**, *1*, 38. [CrossRef]
27. Redyes-Bonilla, H.; López-Pérez, R.A. Corals and coral-reef communities in the Gulf of California. In *Atlas of Coastal Ecosystems in the Western Gulf of California*; Johnson, M.E., Ledesma-Vazquez, J., Eds.; University Arizona Press: Tucson, AZ, USA, 2009; pp. 43–57. ISBN 978-0-8165-2530-0.
28. Merrifield, M.A.; Winant, C.D. Shelf-circulation in the Gulf of California: A description of the variability. *J. Geophys. Res.* **1989**, *94*, 133–160. [CrossRef]
29. Avila, L. The 2015 eastern North Pacific hurricane season: A very active year. *Weatherwise* **2016**, *69*, 36–42. [CrossRef]
30. Travis, C. Earliest Tropical Storm on Record Forms in East Pacific. Available online: https://www.accuweather.com (accessed on 19 May 2021).

*Article*

# Comparison of Modern and Pleistocene (MIS 5e) Coastal Boulder Deposits from Santa Maria Island (Azores Archipelago, NE Atlantic Ocean)

**Sérgio P. Ávila** [1,2,3,4,*], **Markes E. Johnson** [5], **Ana Cristina Rebelo** [1,3,6,7], **Lara Baptista** [1,3,4] **and Carlos S. Melo** [1,3,8,9]

[1] CIBIO, Centro de Investigação em Biodiversidade e Recursos Genéticos, InBIO Laboratório Associado, 9501-801 Ponta Delgada, Portugal; anacrisrebelo@hotmail.com (A.C.R.); laracbaptista@hotmail.com (L.B.); casm.azores@gmail.com (C.S.M.)

[2] Departamento de Biologia, Faculdade de Ciências e Tecnologia da Universidade dos Açores, 9501-801 Ponta Delgada, Portugal

[3] MPB-Marine Palaeontology and Biogeography lab, University of the Azores, 9501-801 Ponta Delgada, Portugal

[4] Faculdade de Ciências da Universidade do Porto, 4169-007 Porto, Portugal

[5] Department of Geosciences, Williams College, Williamstown, MA 01267, USA; mjohnson@williams.edu

[6] Divisão de Geologia Marinha, Instituto Hidrográfico, Rua das Trinas 49, 1249-093 Lisboa, Portugal

[7] Staatliches Museum für Naturkunde Stuttgart, Rosenstein 1, 70191 Stuttgart, Germany

[8] Departamento de Geologia, Faculdade de Ciências, Universidade de Lisboa, 1749-016 Lisboa, Portugal

[9] Instituto Dom Luiz, Faculdade de Ciências, Universidade de Lisboa, 1749-016 Lisboa, Portugal

* Correspondence: avila@uac.pt

Received: 21 April 2020; Accepted: 26 May 2020; Published: 28 May 2020

**Abstract:** Modern and palaeo-shores from Pleistocene Marine Isotope Substage 5e (MIS 5e) featuring prominent cobble/boulder deposits from three locations, on the southern and eastern coast of Santa Maria Island in the Azores Archipelago, were compared, in order to test the idea of higher storminess during the Last Interglacial. A total of 175 basalt clasts from seven transects were measured manually in three dimensions perpendicular to one another. Boulders that exceeded the minimum definitional diameter of 25 cm contributed to 45% of the clasts, with the remainder falling into the category of large cobbles. These were sorted for variations in shape, size, and weight pertinent to the application of two mathematical formulas to estimate wave heights necessary for traction. Both equations were based on the "Nott-Approach", one of them being sensitive to the longest axis, the other to the shortest axis. The preponderance of data derived from the Pleistocene deposits, which included an intertidal invertebrate fauna for accurate dating. The island's east coast at Ponta do Cedro lacked a modern boulder beach due to steep rocky shores, whereas raised Pleistocene palaeo-shores along the same coast reflect surged from an average wave height of 5.6 m and 6.5 m. Direct comparison between modern and Pleistocene deposits at Ponta do Castelo to the southeast and Prainha on the island's south shore produced contrasting results, with higher wave heights during MIS 5e at Ponta do Castelo and higher wave heights for the modern boulder beach at Prainha. Thus, our results did not yield a clear conclusion about higher storminess during the Last Interglacial compared to the present day. Historical meteorological records pit the seasonal activity of winter storms arriving from the WNW-NW against the scant record of hurricanes arriving from the ESE-SE. The disparity in the width of the marine shelf around Santa Maria Island with broad shelves to the north and narrow shelves to the south and east suggested that periodic winter storms had a more regular role in coastal erosion, whereas the rare episodic recurrence of hurricanes had a greater impact on southern and southeastern rocky shores, where the studied coastal boulder deposits were located.

*J. Mar. Sci. Eng.* **2020**, *8*, 386

**Keywords:** coastal boulder deposits; storm surge; hydrodynamic equations; Holocene; Pleistocene; MIS 5e (Marine Isotope Substage 5e); NE Atlantic Ocean

## 1. Introduction

Survey models regarding the level of storm intensity during the Last Interglacial stage of the Pleistocene, specifically the Marine Isotope Substage 5e (MIS 5e), have been conducted on a global scale [1], but studies organized on a more regional scale provide the potential for higher resolution. Localized studies on the propensity for storms in the Bahamas and Bermuda (Western Atlantic) [2,3] reveal patterns in agreement with global results. However, for higher latitudes in the North Atlantic Ocean (e.g., the Azores Archipelago), such analyses have been scarce [4]. Climatological reconstructions are of utmost importance because they allow the scientific community, policymakers, and the general public to better predict and plan for future hazardous events. With a worldwide extension of 500,000 km, coastlines are highly complex and dynamic geomorphological features [5] that commonly correspond to areas of high-density human habitation [6]. With ongoing conditions of global warming, there is even more urgency for increased knowledge about the deep history of storm patterns associated with oceanic circulation.

The shorelines of volcanic oceanic islands are highly dynamic in nature as a result of volcanism, mass wasting, and exposure to the energetic action of the open sea [7–10]. Exposure to wave action is the primary factor that shapes coastlines, as wave surge crossing island shelves acts continuously, but at different levels of energy [11]. In the Azores Archipelago, all islands are subject to the direct action of waves [11]. Unprotected by the absence of barriers (e.g., reefs) and with narrow insular shelves [12], erosion rates are high [6]. The result is the production of variable amounts of detrital materials that are readily shaped and transported, commonly including coastal boulder deposits (CBDs). Sometimes, transported materials are configured in peculiar geomorphologies, such as fajãs [13]. In the Azores, processes that lead to the production of modern CBDs allow for comparisons with morphologies deposited during the Last Pleistocene Interglacial episode (MIS 5e). Santa Maria is the only island in the Azores with marine fossiliferous sequences that date back to the Pliocene and late Pleistocene (MIS 5e) [14–16]. The presence of such sequences makes this island an ideal place for studies testing the postulated higher storminess and inferred palaeo-wave heights that affected the wider archipelago approximately 125,000 years ago.

Inference on palaeo-wave-heights through measurements of eroded blocks in Pleistocene settings has been conducted by Johnson et al. [17–19] at localities in the Mexican Gulf of California, providing a useful methodology for application elsewhere. Here, we adapted the program using mathematical formulas to compare the storminess during the Last Interglacial (MIS 5e) and the modern situation, deduced from storms imprinted in the CBDs. Herein, we presented the first estimations for palaeo-wave heights from the Pleistocene (MIS 5e) in the Azores. Data on the shape, size, and calculated weight of palaeo-shore boulders from Santa Maria Island had the advantage of being sourced from well-dated MIS 5e outcrops containing thermophile fauna typical of the Last Interglacial. Crucially, estimates on wave heights from MIS 5e CBDs might be compared with adjacent modern beach boulders. Work was limited to localities on the island's south and east coast, where marine shelves were narrower. The island's broader northern shelf provided the basis for conjecture on differences in storm patterns that led to that outcome.

## 2. Geographical and Geological Setting

### 2.1. Position and Geotectonic Setting

The Azores is an oceanic volcanic archipelago with nine islands, found in the NE North Atlantic Ocean, between latitudes 36°55′ and 49°43′ N and longitudes 24°46′ and 31°16′ W, and spreading along a distance of 650 km (Figure 1a). Seated on the Azores Plateau, the archipelago straddles a triple junction between the Eurasian, North American, and Nubian (African) tectonic plates [20]. Santa Maria Island is the most southeastern and oldest among these and is seated on the Nubian plate. It exhibits a geologic record, marking the earliest emergence from the sea due to Surtseyan volcanic activity approximately 6 million years ago [8]. Much of the ensuing rock record, which includes intercalated sedimentary strata and extensive volcanic flows, is restricted to the Pliocene Epoch [21,22]. Island uplift, which commenced 2.8 million years ago, has resulted in more than 200 m of tectonic rise, as a result of which older Pliocene marine strata are well exposed in sea cliffs on all sides of the island, but most accessible along the south and eastern shores [23]. Santa Maria ranks seventh in size compared to the other islands in the Azores Archipelago, with an area of 97 km$^2$ and a coastal circumference of 53 km. The island presents a peculiar orography, with a flatter Western part and a rougher Eastern part, as a result of its unusual geological evolution (i.e., different erosional rates and off-center volcanism during a rejuvenated stage, mostly located on the eastern section of the older edifice [8,24]; Figure 1b). In terms of bathymetry, the island is much reduced in size with an asymmetrical marine shelf that is broadest to the north and narrowest to the south and east [24]. The shelf is also characterized by a suite of submerged terraces, all presumably younger than ~1 Ma, which are more developed and preserved in wider and low-gradient sectors (Figure 1b) [25]. During the Last Interglacial epoch of the Pleistocene, extensive boulder beds were emplaced all around the island that incorporates marine fossils attributed to the MIS 5e [4]. Three localities, including Prainha and Ponta do Castelo on the south shore and Ponta do Cedro on the east coast, were chosen to highlight lateral variations in Pleistocene boulder size together with a review of the associated fauna. Comparisons were drawn with modern boulder beds at the two southern localities, whereas the plunging coastal cliffs at Ponta do Cedro lacked any such development of a coastal boulder deposit (CBD) at or close to modern sea level.

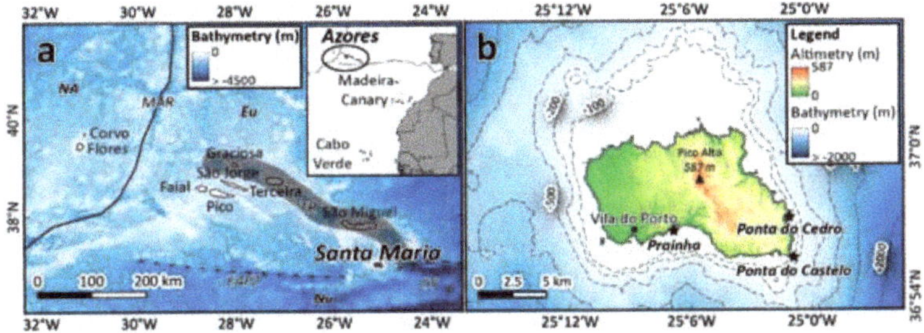

**Figure 1.** Maps covering the Azores Archipelago in the NE Atlantic Ocean: (**a**) Geographic and geotectonic setting of the Azores Archipelago (modified from [26–28]). NA—North American plate; Eu—Eurasian plate; Nu—Nubian (African) plate; MAR – Mid Atlantic Ridge; TR—Terceira Rift (grey area); EAFZ—East Azores Fracture Zone; GF—Gloria Fault; (**b**) Topography/bathymetry of Santa Maria Island. Black stars mark the studied sites. The bathymetric data was extracted from GEBCO 2019 (https://www.gebco.net); subaerial topography was generated from a 1:5000 scale digital altimetric database.

## 2.2. Sources and Shaping of the Boulders

The Prainha site is situated within a bay, thus affording a higher potential for the deposition of clasts and finer sediments. The analysis of local currents showed a confluence of waves to the south coast of the island (https://www.ipma.pt/pt/maritima/hs/index.jsp?area=acores-east) that explains how during the summer season the nearby coastlines collect extensive (on a local scale) sand beaches. The area is also the river-mouth of some streams. As the archipelago's climate is classified as "Warm Temperate" [29], higher precipitation events are common. Such events are responsible for the transport of varying amounts of boulders and sand to the intertidal zone. As rough seas are commonly registered in the archipelago, the boulders are easily rounded and shaped.

At the southeasternmost tip of the island, Ponta do Castelo presents the most complex wave regime from the three sites studied. In the area, perpendicular (in direction) wave currents interact, promoting rougher seas, even under fair-weather conditions. As a result of this exposure, no fine sediment is deposited, and only a well-developed CBD is present.

In Ponta do Cedro, however, despite being found inside a bay, like Prainha, a plunging coastline allows for no accumulation of sandy deposits, and a modern CBD also is absent. Ponta do Cedro and Ponta do Castelo, however, share the same type of boulder source resulting from erosion of the area's high sea cliffs. Both these sites are situated in the rough eastern end of the island, where sea cliffs are impressive in height.

### 2.3. Wave Energy and Direction

The Azores occupies a region in the NE Atlantic Ocean encompassing approximately 22,500 km$^2$ (Figure 1a) and is characterized by a high level of marine storm activity. Hurricanes rarely cross through the region during the annual Atlantic hurricane season, although tropical cyclones are common. Such storms arrive from a zone off the west coast of Africa in the vicinity of the Cabo Verde Archipelago, located at a lower 17° N latitude. Notwithstanding, extreme storm events tend to arrive once every seven years, on average [6,30]. In contrast and as a result of the exposure to the strong NW North Atlantic winds – the Westerlies [31], the northern exposures of islands are subject to winter storms. The archipelago is also in the pathway the Gulf Stream makes from the North American coasts towards the central zone of the North Atlantic [32,33], being so a source of many instability processes, meanders, and eddies [34]. The main wave direction that affects the archipelago comes from WNW-NW [5,8,23]. Such predominance is reflected in the island's bathymetrical morphology that shows higher erosional rates on the W and N coasts. According to Rusu and Soares [11], and using the closest station to Santa Maria Island (P18), present-day mean wave heights during summer is 1.75 and during winter is 3.20 m. For the Azores archipelago, Santa Maria presents the lowest values regarding the mean values for both seasons, and Corvo presents the highest. This variation in values has been related to the shadow effect promoted mainly by São Miguel Island (North of Santa Maria), protecting it from the high wave energy [13].

## 3. Methods

### 3.1. Data Collection

Original data were collected for both modern and MIS 5e coastal boulder deposits at two localities (Prainha and Ponta do Castelo, Figure 1b) and only MIS 5e from one locality (Ponta do Cedro, Figure 1b). Measurements from modern deposits were performed at mean sea level, while Pleistocene deposit heights varied from ~2.5 m and as much as ~9.0 m above mean sea level (amsl) at five localities, all on the southern and eastern shores of Santa Maria Island. The definition for a boulder adapted to this exercise is that of Wentworth [35] for an erosional clast equal or greater than 256 mm and less than 4096 mm in diameter, with cobbles defined as clasts pertaining to the class 64-256 mm in maximum diameter. For each data set, 25 of the largest available clasts were selected along a transect line parallel to the shore spaced no more than one meter apart. Each clast required three measurements along principle axes (long *a*, intermediate *b*, and short *c*). The initial calculation for volume was simply a multiplication of the long, intermediate, and short-axis values. In all cases, this resulted in a cubical to a rectangular shape that did not take into account the rounding of the boulders by erosion; therefore, a final adjustment to 75% was made, regarding the volume estimates for each boulder, in accordance with previous works [18,19]. Triangular plots were employed to demonstrate variations in boulder shape, following the practice of Sneed and Folk for river pebbles [36]. Data regarding the maximum and intermediate lengths perpendicular to one another from individual boulders were plotted in bar graphs to show potential shifts in size from one transect to the next. A brief description of the fossil fauna in each site was also provided. This was important information, as, together with the geological data, the presence of the characteristic MIS 5e thermophilic taxa [4] attested to the age of the deposit, which was not possible to date due to the absence of suitable biogenic material (e.g., corals).

*3.2. Hydraulic Model*

With the determination of specific gravity based on the standard density value for oceanic basalt at 3.0 g/cm$^3$, a hydraulic model might be applied to predict the energy needed to transfer larger basalt blocks from rocky shoreline to an adjacent coastal boulder deposit as a function of wave impact. Basalt is a volcanic rock that forms from surface flows with variable thicknesses and a propensity to develop vertical fractures. These factors regulate the size and general shape of blocks loosened in the cliff face [6]. Herein, we used two different formulas to estimate the magnitude of storm waves applied to joint-bounded boulders derived, respectively, from Equation (36) in the work of Nott [37] (Equation (1)) and from a recent formula of Pepe et al. [38] that used the velocity equations of Nandasena et al. [39] to estimate wave heights (Equation (2)):

$$H_S = \frac{\left(\frac{\rho_s - \rho_w}{\rho_w}\right)a}{C_l} \tag{1}$$

$$H_S = \frac{\left(\frac{\left(\frac{\rho_s - \rho_w}{\rho_w}\right)}{C_l}\right) \cdot c \cdot \left(\frac{\cos\theta + \mu_s \times \sin\theta}{C_l}\right)}{100} \tag{2}$$

where $H_s$ is the height of the storm wave at breaking point; $\rho_s$ is the density of the boulder (3 tons/m$^3$ or 3 g/cm$^3$); $\rho_w$ is the density of water at 1.02 g/cm$^3$; $a$ is the length of the boulder on long axis in cm; $c$ is the length of the boulder on short axis in cm; $\theta$ is the angle of the bed slope at the pre-transport location (1° for joint-bounded boulders); $\mu_s$ is the coefficient of static friction (=0.7); $C_l$ is the lift coefficient (=0.178). Equation (1) was more sensitive to the length of a boulder at the long axis, whereas Equation (2) was more sensitive to the length of a boulder on the short axis. Therefore, some differences were expected in the estimates of $H_S$.

## 4. Results

*4.1. Prainha on the South Shore*

Located 3.5 km east of the harbor at Vila do Porto (Figure 1b), Prainha's area presented well-developed modern CBD, at heights of ~0.5 m amsl (Figure 2), as well as exposed Pleistocene marine sequences, deposited on top of volcaniclastic rocks at an elevation of ~3 m amsl [40] (Figure 3). Although CBDs were present all-year-round at Prainha, the location of this site (south coast of the island) combined with the fact that Prainha was found within a bay, and taking into consideration the wave regime, it resulted that, in some cases, the modern CBD was covered by sand. Raw data on clast size in three dimensions collected from the two parallel transects at this locality are present in Tables 1 and 2.

Data points representing individual boulders grouped by transect were plotted on a set of Sneed-Folk triangular diagrams (Figure 4a,b), showing shape variations. Those points clustered nearest to the core of the diagrams were most faithful to an average value with somewhat equidimensional axes in three directions. Only rarely any points fell into the upper-most triangle, which signified a cube-shaped endpoint.

The majority of points from both sets fell within the central part of the two tiers beneath the top triangle. However, the overall trend shared between the two sets traced a similar pattern angled toward the lower right corner of the diagrams. The modern CBD at Prainha (Figures 2 and 4a) demonstrated a greater tendency to elongate shapes represented by the endpoint for bar-shaped clasts. No points appeared in the lower-left tier of either diagram, which represented an endpoint reserved for plate-shaped clasts. Although the general trend in slope was similar between the modern and Pleistocene CBDs, the plots had no bearing on actual variations in clast size.

Variations in boulder size as a function of maximum and intermediate axis length were plotted using bar graphs (Figure 5), based on raw data drawn from Table 1. The greatest number of boulders in the sample measured from the modern CBD fell within a maximum diameter size range between 26 and 35 cm (Figure 5a), which qualified as small boulders in the Wentworth scheme [35]. The largest boulders encountered at Prainha were few in number but ranged in size between 46 and 55 cm. The tendency towards an elongated shape among these clasts was shown by a marked shift in the dominant bin-size for the intermediate axis, within an interval of 16 to 25 cm (Figure 5b).

**Figure 2.** Modern coastal boulder deposit (CBD) eroded from adjacent basalt sea cliffs at Prainha (site 1). Note that some boulders on the far inland end of the littoral boulder cordon were embedded within the sand.

**Figure 3.** Pleistocene (MIS 5e, Marine Isotope Substage 5e) marine sequence, 2.5 m above the modern CBD at Prainha (site 2). Note that, as modern CBD, Pleistocene boulders were covered by Pleistocene fossiliferous sands.

**Table 1.** Quantification of boulder size, volume, and estimated weight from the modern coastal boulder deposit at site 1 at Prainha, on the southern coast of Santa Maria Island (see Figure 1b). The standard density of basalt at 3.0 gm/cm$^3$ was applied uniformly in order to calculate wave height for each boulder. EWH: estimated wave height (in meters), calculated according to equations from Nott [37] and Pepe et al. [38]. See the methods Section 3.2. (hydraulic model).

| Sample | Long Axis (cm) | Intermediate Axis (cm) | Short Axis (cm) | Volume (cm$^3$) | Adjust. to 75% | Weight (kg) | EWH Nott [37] | EWH Pepe et al. [38] |
|---|---|---|---|---|---|---|---|---|
| 1 | 28.0 | 24.5 | 6.0 | 4116 | 3087 | 9.3 | 3.1 | 1.5 |
| 2 | 32.0 | 24.0 | 8.0 | 6144 | 4608 | 13.8 | 3.5 | 2.0 |
| 3 | 30.0 | 23.0 | 10.5 | 7245 | 5434 | 16.3 | 3.3 | 2.6 |
| 4 | 26.0 | 17.5 | 8.0 | 3640 | 2730 | 8.2 | 2.8 | 2.0 |
| 5 | 24.5 | 18.5 | 9.0 | 4079 | 3059 | 9.2 | 2.7 | 2.2 |
| 6 | 28.0 | 24.5 | 8.0 | 5488 | 4116 | 12.3 | 3.1 | 2.0 |
| 7 | 48.0 | 23.0 | 19.0 | 20,976 | 15,732 | 47.2 | 5.2 | 4.7 |
| 8 | 44.0 | 37.0 | 8.0 | 13,024 | 9768 | 29.3 | 4.8 | 2.0 |
| 9 | 43.0 | 29.0 | 8.0 | 9976 | 7482 | 22.4 | 4.7 | 2.0 |
| 10 | 38.0 | 25.0 | 7.5 | 7125 | 5344 | 16.0 | 4.1 | 1.8 |
| 11 | 39.5 | 27.0 | 11.0 | 11,732 | 8799 | 26.4 | 4.3 | 2.7 |
| 12 | 31.0 | 23.0 | 6.5 | 4635 | 3476 | 10.4 | 3.4 | 1.6 |
| 13 | 23.0 | 18.5 | 8.5 | 3617 | 2713 | 8.1 | 2.5 | 2.1 |
| 14 | 31.0 | 28.0 | 12.0 | 10,416 | 7812 | 23.4 | 3.4 | 3.0 |
| 15 | 28.0 | 20.5 | 9.0 | 5166 | 3875 | 11.6 | 3.1 | 2.2 |
| 16 | 34.5 | 25.5 | 13.0 | 11,437 | 8578 | 25.7 | 3.8 | 3.2 |
| 17 | 32.0 | 26.0 | 9.0 | 7488 | 5616 | 16.8 | 3.5 | 2.2 |
| 18 | 26.0 | 17.0 | 16.0 | 7072 | 5304 | 15.9 | 2.8 | 3.9 |
| 19 | 24.5 | 19.0 | 8.0 | 3724 | 2793 | 8.4 | 2.7 | 2.0 |
| 20 | 43.0 | 27.0 | 26.0 | 30,186 | 22,640 | 67.9 | 4.7 | 6.4 |
| 21 | 42.0 | 34.0 | 16.0 | 22,848 | 17,136 | 51.4 | 4.6 | 3.90 |
| 22 | 32.0 | 23.0 | 22.0 | 16,192 | 12,144 | 36.4 | 3.5 | 5.4 |
| 23 | 43.5 | 22.5 | 20.0 | 19,575 | 14,681 | 44.0 | 4.7 | 4.9 |
| 24 | 31.0 | 24.0 | 17.0 | 12,648 | 9486 | 28.5 | 3.4 | 4.2 |
| 25 | 13.0 | 10.0 | 8.5 | 1105 | 829 | 2.5 | 1.4 | 2.1 |

**Table 2.** Quantification of boulder size, volume, and estimated weight from the Pleistocene (MIS 5e, Marine Isotope Substage 5e) coastal conglomerate at site 2 at Prainha, on the southern coast of Santa Maria Island (see Figure 1b). The standard density of basalt at 3.0 gm/cm$^3$ was applied uniformly in order to calculate wave height for each boulder. EWH: estimated wave height (in meters), calculated according to equations from Nott [37] and Pepe et al. [38]. See the methods Section 3.2. (hydraulic model).

| Sample | Long Axis (cm) | Intermediate Axis (cm) | Short Axis (cm) | Volume (cm$^3$) | Adjust. to 75% | Weight (kg) | EWH Nott, [37] | EWH Pepe et al. [38] |
|---|---|---|---|---|---|---|---|---|
| 1 | 12.5 | 11.0 | 8.5 | 1169 | 877 | 2.6 | 1.4 | 2.1 |
| 2 | 8.5 | 6.0 | 4.0 | 204 | 153 | 0.5 | 0.9 | 1.0 |
| 3 | 8.0 | 5.0 | 4.0 | 160 | 120 | 0.4 | 0.9 | 1.0 |
| 4 | 12.0 | 8.5 | 6.5 | 663 | 497 | 1.5 | 1.3 | 1.6 |
| 5 | 10.5 | 6.5 | 4.0 | 273 | 205 | 0.6 | 1.1 | 1.0 |
| 6 | 13.0 | 11.0 | 10.5 | 1502 | 1126 | 3.4 | 1.4 | 2.6 |
| 7 | 9.0 | 8.0 | 3.0 | 216 | 162 | 0.5 | 1.0 | 0.7 |
| 8 | 25.0 | 13.5 | 13.0 | 4388 | 3291 | 9.9 | 2.7 | 3.2 |
| 9 | 10.0 | 7.0 | 6.0 | 420 | 315 | 0.9 | 1.1 | 1.5 |
| 10 | 8.0 | 5.0 | 4.0 | 160 | 120 | 0.4 | 0.9 | 1.0 |
| 11 | 9.0 | 5.0 | 4.5 | 203 | 152 | 0.5 | 1.0 | 1.1 |
| 12 | 10.0 | 8.0 | 6.0 | 480 | 360 | 1.1 | 1.1 | 1.5 |
| 13 | 11.0 | 7.0 | 4.0 | 308 | 231 | 0.7 | 1.2 | 1.0 |
| 14 | 16.5 | 10.5 | 5.0 | 866 | 650 | 1.9 | 1.8 | 1.2 |
| 15 | 8.5 | 5.5 | 2.0 | 94 | 70 | 0.2 | 0.9 | 0.5 |
| 16 | 15.0 | 6.0 | 5.0 | 450 | 338 | 1.0 | 1.6 | 1.2 |
| 17 | 14.5 | 7.5 | 5.0 | 544 | 408 | 1.2 | 1.6 | 1.2 |
| 18 | 8.5 | 5.5 | 2.5 | 117 | 88 | 0.3 | 0.9 | 0.6 |
| 19 | 9.0 | 5.5 | 4.5 | 223 | 167 | 0.5 | 1.0 | 1.1 |
| 20 | 10.5 | 8.0 | 4.5 | 378 | 284 | 0.9 | 1.1 | 1.1 |
| 21 | 7.0 | 5.5 | 3.0 | 116 | 87 | 0.3 | 0.8 | 0.7 |
| 22 | 12.0 | 7.5 | 5.5 | 495 | 371 | 1.1 | 1.3 | 1.4 |
| 23 | 9.0 | 5.5 | 3.0 | 149 | 111 | 0.3 | 1.0 | 0.7 |
| 24 | 14.5 | 8.5 | 4.0 | 493 | 370 | 1.1 | 1.6 | 1.0 |
| 25 | 20.5 | 11.0 | 8.0 | 1804 | 1353 | 4.1 | 2.2 | 2.0 |

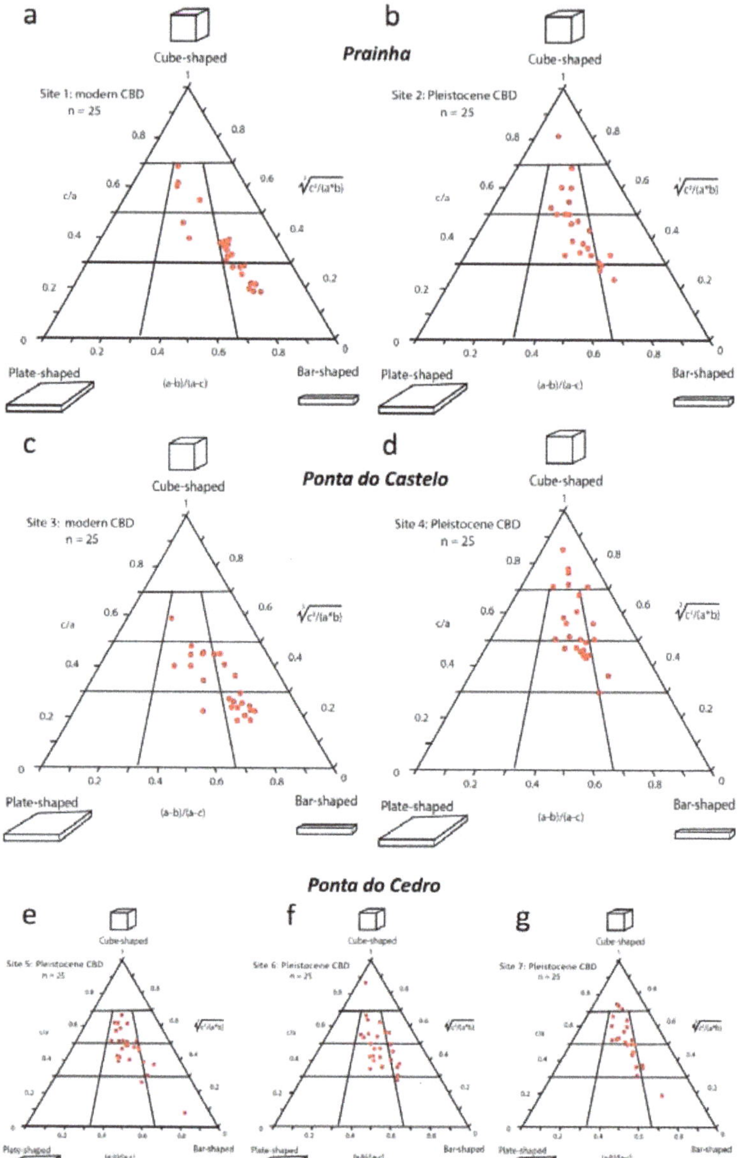

**Figure 4.** Set of triangular Sneed-Folk diagrams used to show variations in boulder and cobble shapes compared between modern and Pleistocene (MIS 5e) deposits. (**a,b**): Prainha. (**c,d**): Ponta do Castelo. (**e–g**): Ponta do Cedro. (**a,c**): modern CBD at Prainha and Ponta do Castelo, respectively. (**b,d**), (**e–g**): Pleistocene (MIS 5e) CBD at Prainha, Ponta do Castelo, and Ponta do Cedro, respectively.

In strong contrast, all clasts encountered from the Pleistocene (MIS 5e) conglomerate fell into the category of cobbles as defined in the Wentworth scheme [35]. By far, the largest number of clasts within the sample fell into the size range between 6 and 15 cm in maximum diameter (Figure 5c), also replicated by the same frequency for the intermediate axis (Figure 5d). However, a subsample of smaller clasts in the size range of pebbles to small cobbles appeared in the sample measured for the intermediate axis. Comparing the modern CBD at Prainha (Figure 2) with the general contents of the Pleistocene conglomerate at the same locality, it was clear that wave conditions on the modern shore eroded significantly larger clasts from the parent basalt rocky shore.

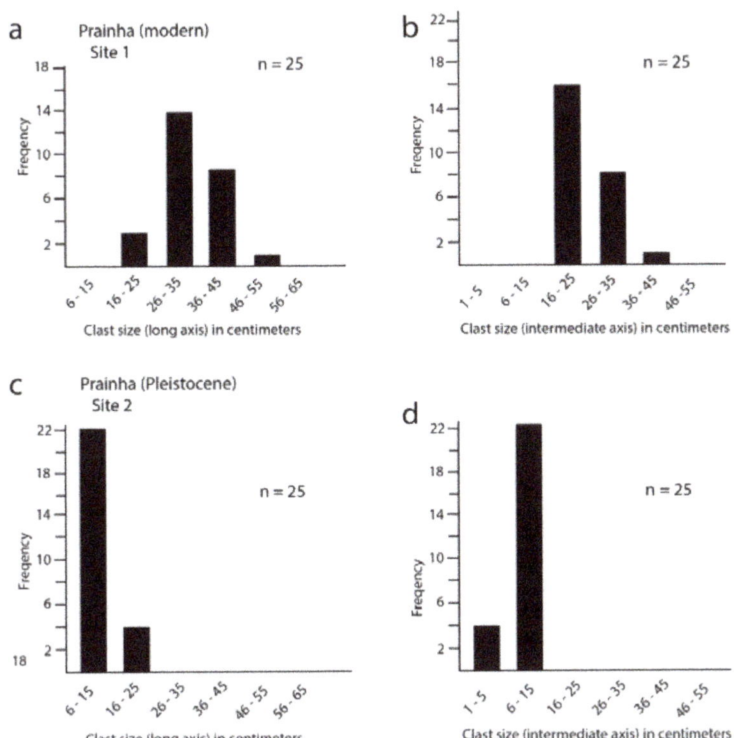

**Figure 5.** Bar graphs used to appraise variations in the long and intermediate axes on basalt clasts from modern and Pleistocene CBDs at Prainha; (**a**) Long-axis length from clasts in the modern CBD; (**b**) Intermediate-axis length from clasts in the modern CBD; (**c**) Long-axis length from clasts in the Pleistocene CBD; (**d**) Intermediate-axis length from clasts in the Pleistocene CBD.

### 4.2. Context of the Pleistocene Fauna at Prainha

Prainha is one of the six Pleistocene (MIS 5e) fossiliferous outcrops described for Santa Maria Island. MIS 5e CBDs are known from Prainha, Ponta do Castelo, Ponta do Cedro, and Pedra-que-pica, but do not occur at Lagoinhas nor at Vinha Velha.

Prainha is the best-studied outcrop from Santa Maria Island, and its fossils being first reported by Portuguese geologists [41,42]. After these pioneer works, Santa Maria outcrops' and their fossiliferous remains have been systematically described, with a total of 148 fossil marine-specific taxa presently reported from the Santa Maria Last Interglacial deposits [4]. The most biodiverse invertebrate group is, by far, the marine mollusks, with a total of 138 taxa (114 Gastropoda and 24 Bivalvia) [41,43–45] reported for all MIS 5e deposits in the island (of these, 100 gastropods and 20 bivalve taxa are reported from Prainha). It is followed by the Echinodermata (three taxa) and Cnidaria Anthozoa (one coral taxa) [40,46,47]. Four species of coralline red algae have also been reported from the MIS 5e of Prainha [48]. Finally, rare vertebrate remains have been described from the MIS 5e sedimentary sequences in Santa Maria: one bony fish species, *Sparisoma cretense* (Linnaeus, 1758) reported from the MIS 5e of Vinha Velha [49], and one undetermined Balaenopteridae species (a baleen whale) reported from the MIS 5e of Prainha [50]. As a result of its high palaeobiodiversity, a high number of scientific studies, education and touristic potential, and the presence of extremely rare vertebrate cetacean remains, Prainha is considered as a high-relevance, national geosite [15,16].

*4.3. Ponta Do Castelo at the Island's Southeast End*

Located approximately 12 km farther east from Prainha (Figure 1b), the study site at Ponta do Castelo occupied the extreme southeast corner of Santa Maria Island. There, a modern CBD was entrained as a berm at mean sea level representing site 3, whereas the Pleistocene (MIS 5e) CBD was lodged above Pliocene strata at a height ~4.3 m amsl, representing site 4 (Figure 6).

**Figure 6.** Modern coastal boulder deposit (CBD) eroded from basalt at Ponta do Castelo (site 3, see map in Figure 1b) and overlying Pleistocene CBD preserved at a level 4.3 m above mean sea level at the same locality (sites 4 and 5).

The high exposure of this site to the main sea currents and the local morphology did not allow the deposition and maintenance of sand deposits. Raw data on clast size in three dimensions collected from the two parallel transects at this locality are available in Tables 3 and 4.

Data points representing individual boulders grouped by transect were plotted on a set of Sneed-Folk triangular diagrams (Figure 4c,d), showing shape variations. Compared to the pair of Sneed-Folk diagrams from Prainha (Figure 4a,b), the Ponta do Castelo set was similar with regard to the slope of points angled to the lower-right corner (Figure 4c,d). The main difference was that the data set from the modern CBD at Ponta do Castelo (Figure 4c) was more diffuse throughout the same subdivisions represented by the modern CBD at Prainha (Figure 4a). A higher number of points registered in the top triangle marked a significant deviation in the Pleistocene (MIS 5e) CBD at Ponta do Castelo (Figure 4d) compared with Prainha (Figure 4b). Overall, there was a greater tendency towards equidimensional clasts in the Pleistocene (MIS 5e) CBD than in the modern CBD at Ponta do Castelo.

**Table 3.** Quantification of boulder size, volume, and estimated weight from the modern coastal boulder deposit at site 3 at Ponta do Castelo, from the southeast end of Santa Maria Island (see map, Figure 1b). The standard density of basalt at 3.0 gm/cm$^3$ was applied uniformly in order to calculate wave height for each boulder. EWH: estimated wave height (in meters), calculated from equations in Nott [37] and Pepe et al. [38]. See the methods Section 3.2. (hydraulic model).

| Sample | Long Axis (cm) | Intermediate Axis (cm) | Short Axis (cm) | Volume (cm$^3$) | Adjust. to 75% | Weight (kg) | EWH Nott, [37] | EWH Pepe et al. [38] |
|---|---|---|---|---|---|---|---|---|
| 1 | 50.0 | 15.0 | 11.0 | 8250 | 6188 | 18.6 | 5.5 | 2.7 |
| 2 | 35.0 | 14.0 | 14.0 | 6860 | 5145 | 15.4 | 3.8 | 3.4 |
| 3 | 25.0 | 22.0 | 6.0 | 3300 | 2475 | 7.4 | 2.7 | 1.5 |
| 4 | 27.5 | 13.0 | 5.0 | 1788 | 1341 | 4.0 | 3.0 | 1.2 |
| 5 | 18.0 | 15.0 | 4.0 | 1080 | 810 | 2.4 | 2.0 | 1.0 |
| 6 | 23.5 | 19.0 | 6.0 | 2679 | 2009 | 6.0 | 2.6 | 1.5 |
| 7 | 22.5 | 20.0 | 10.0 | 4500 | 3375 | 10.1 | 2.5 | 2.5 |
| 8 | 32.0 | 29.5 | 7.0 | 6608 | 4956 | 14.9 | 3.5 | 1.7 |
| 9 | 32.0 | 17.0 | 11.0 | 5984 | 4488 | 13.5 | 3.5 | 2.7 |
| 10 | 31.0 | 16.0 | 12.5 | 6200 | 4650 | 14.0 | 3.4 | 3.1 |
| 11 | 25.0 | 16.0 | 5.0 | 2000 | 1500 | 4.5 | 2.7 | 1.2 |
| 12 | 22.0 | 20.0 | 9.0 | 3960 | 2970 | 8.9 | 2.4 | 2.2 |
| 13 | 27.0 | 18.5 | 5.0 | 2498 | 1873 | 5.6 | 2.9 | 1.2 |
| 14 | 20.0 | 14.0 | 9.0 | 2520 | 1890 | 5.7 | 2.2 | 2.2 |
| 15 | 23.0 | 15.5 | 6.0 | 2139 | 1604 | 4.8 | 2.5 | 1.5 |
| 16 | 21.0 | 13.5 | 4.3 | 1219 | 914 | 2.7 | 2.3 | 1.1 |
| 17 | 22.0 | 20.0 | 9.0 | 3960 | 2970 | 8.9 | 2.4 | 2.2 |
| 18 | 18.0 | 17.5 | 6.5 | 2048 | 1536 | 4.6 | 2.0 | 1.6 |
| 19 | 26.0 | 16.5 | 12.5 | 5363 | 4022 | 12.1 | 2.8 | 3.1 |
| 20 | 24.0 | 21.0 | 7.0 | 3528 | 2646 | 7.9 | 2.6 | 1.7 |
| 21 | 25.5 | 15.0 | 15.0 | 5738 | 4303 | 12.9 | 2.8 | 3.7 |
| 22 | 19.0 | 12.5 | 4.5 | 1069 | 802 | 2.4 | 2.1 | 1.1 |
| 23 | 21.0 | 12.5 | 5.0 | 1313 | 984 | 3.0 | 2.3 | 1.2 |
| 24 | 20.5 | 11.5 | 9.0 | 2122 | 1591 | 4.8 | 2.2 | 2.2 |
| 25 | 27.5 | 18.0 | 7.5 | 3713 | 2784 | 8.4 | 3.0 | 1.8 |

**Table 4.** Quantification of boulder size, volume, and estimated weight from the Pleistocene (MIS 5e) conglomerate at site 4 at Ponta do Castelo, from the southeast end of Santa Maria Island (see map, Figure 1b). The standard density of basalt at 3.0 gm/cm$^3$ was applied uniformly in order to calculate wave height for each boulder. EWH: estimated wave height (in meters), calculated from equations in Nott [37] and Pepe et al. [38]. See the methods Section 3.2. (hydraulic model).

| Sample | Long Axis (cm) | Intermediate Axis (cm) | Short Axis (cm) | Volume (cm$^3$) | Adjust. to 75% | Weight (kg) | EWH Nott, [37] | EWH Pepe et al. [38] |
|---|---|---|---|---|---|---|---|---|
| 1 | 39.0 | 35.0 | 33.0 | 45,045 | 33,784 | 101.4 | 4.3 | 8.1 |
| 2 | 49.0 | 32.0 | 25.0 | 39,200 | 29,400 | 88.2 | 5.3 | 6.2 |
| 3 | 17.0 | 15.0 | 6.0 | 1530 | 1148 | 3.4 | 1.9 | 1.5 |
| 4 | 15.0 | 8.0 | 7.5 | 900 | 675 | 2.0 | 1.6 | 1.8 |
| 5 | 32.0 | 30.0 | 16.0 | 15,360 | 11,520 | 34.6 | 3.5 | 3.9 |
| 6 | 22.0 | 15.0 | 10.0 | 3300 | 2475 | 7.4 | 2.4 | 2.5 |
| 7 | 29.0 | 20.0 | 17.0 | 9860 | 7395 | 22.2 | 3.2 | 4.2 |
| 8 | 11.0 | 10.0 | 8.5 | 935 | 701 | 2.1 | 1.2 | 2.1 |
| 9 | 25.0 | 17.0 | 14.0 | 5950 | 4463 | 13.4 | 2.7 | 3.4 |
| 10 | 25.0 | 22.5 | 19.0 | 10,688 | 8016 | 24.0 | 2.7 | 4.7 |
| 11 | 33.0 | 28.0 | 20.0 | 18,480 | 13,860 | 41.6 | 3.6 | 4.9 |
| 12 | 23.0 | 16.0 | 10.0 | 3680 | 2760 | 8.3 | 2.5 | 2.5 |
| 13 | 20.0 | 14.0 | 9.0 | 2520 | 1890 | 5.7 | 2.2 | 2.2 |
| 14 | 30.0 | 20.0 | 14.0 | 8400 | 6300 | 18.9 | 3.3 | 3.4 |
| 15 | 14.0 | 12.0 | 10.0 | 1680 | 1260 | 3.8 | 1.5 | 2.5 |
| 16 | 18.5 | 20.0 | 13.0 | 4810 | 3608 | 10.8 | 2.0 | 3.2 |
| 17 | 14.0 | 10.5 | 7.0 | 1029 | 772 | 2.3 | 1.5 | 1.7 |
| 18 | 32.5 | 18.0 | 15.0 | 8775 | 6581 | 19.7 | 3.5 | 3.7 |
| 19 | 22.5 | 18.0 | 11.0 | 4455 | 3341 | 10.0 | 2.5 | 2.7 |
| 20 | 17.0 | 12.0 | 12.0 | 2448 | 1836 | 5.5 | 1.9 | 3.0 |
| 21 | 15.0 | 14.0 | 10.0 | 2100 | 1575 | 4.7 | 1.6 | 2.5 |
| 22 | 28.0 | 20.0 | 12.0 | 6720 | 5040 | 15.1 | 3.1 | 3.0 |
| 23 | 25.0 | 25.0 | 14.0 | 8750 | 6563 | 19.7 | 2.7 | 3.4 |
| 24 | 35.0 | 20.0 | 10.0 | 7000 | 5250 | 15.8 | 3.8 | 2.5 |
| 25 | 32.0 | 24.0 | 14.0 | 10,752 | 8064 | 24.2 | 3.5 | 3.4 |

At Ponta do Castelo, variations in boulder size as a function of maximum and intermediate axis length were plotted using bar graphs (Figure 7), based on the raw data drawn from Tables 3 and 4. The general congruence between the modern and Pleistocene (MIS 5e) CBDs at Ponta do Castelo was strong, especially compared to the marked difference in size variation observed between the modern and Pleistocene (MIS 5e) CBDs at Prainha (Figure 5). Of foremost significance, 50% of measurements for clast size through the long axis (Figure 7a) qualified as boulders based on the criteria of Wentworth [35]. The tendency towards elongated clasts in the modern CBD was shown by the higher frequency in the bin size 6–15 cm for the intermediate axis (Figure 7b), whereas that bin size was void with respect to the long axis (Figure 7a). Compared to the modern CBD, the Pleistocene CBD at Ponta do Castelo exhibited only a small shift of data points into the size interval of 6 to 15 cm, but otherwise was similar. The congruence between modern and Pleistocene CBDs at Ponta do Castelo was even more striking, taking into account clast size measured on the intermediate axes (Figure 7b,d).

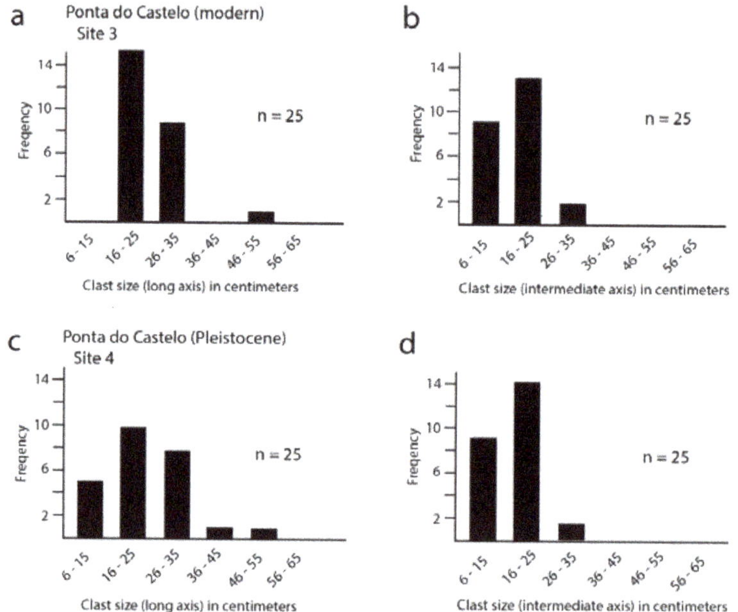


**Figure 7.** Bar graphs used to appraise variations in the long and intermediate axes on basalt clasts from modern and Pleistocene CBDs at Ponta do Castelo; (**a**) Long-axis length from clasts in the modern CBD; (**b**) Intermediate-axis length from clasts in the modern CBD; (**c**) Long-axis length from clasts in the Pleistocene CBD; (**d**) Intermediate-axis length from clasts in the Pleistocene CBD.

### 4.4. Context of the Pleistocene Fauna at Ponta Do Castelo

The MIS 5e sedimentary sequence at Ponta do Castelo is extremely poor, with only two gastropod species reported: the limpet *Patella aspera* Röding, 1798, and the supra-littoral littorinid *Tectarius striatus* (P.P. King, 1832). Nevertheless, Ponta do Castelo is considered as a geosite of international relevance because of a set of specific conditions that allowed the preservation of a shelf tempestite deposit [21] for which a precise water depth could be estimated. Additionally, it provides a good proxy for island uplift/subsidence reconstructions [15,16].

### 4.5. Ponta Do Cedro on the Island's East Shore

Located 3.0 km north of Ponta do Castelo, the third study area treated herein occurred on the east shore of Santa Maria Island (Figure 1b). No modern CBDs occurred at this locality. However, a correlated line of Pleistocene CBDs might be traced laterally at elevations that ranged between

~2.65 m and ~9 m amsl, represented by study sites 5 to 7. Typically, the conglomerate at Ponta do Cedro was as much as ~0.75 m in thickness and laterally continuous (Figure 8, site 6).

**Figure 8.** Pleistocene (MIS 5e) conglomerate 9 m above mean sea level at Ponta do Cedro (site 6).

Raw data on clast size in three dimensions collected from three study sites at this locality are available in Tables 5–7.

**Table 5.** Quantification of boulder size, volume, and estimated weight from the Pleistocene (MIS 5e) conglomerate at site 5 at Ponta do Cedro, on the east shore of Santa Maria Island (see map, Figure 1b). The standard density of basalt at 3.0 gm/cm$^3$ was applied uniformly in order to calculate wave height for each boulder. EWH: estimated wave height (in meters), calculated from equations in Nott [37] and Pepe et al. [38]. See the methods Section 3.2. (hydraulic model).

| Sample | Long Axis (cm) | Intermediate Axis (cm) | Short Axis (cm) | Volume (cm$^3$) | Adjust. to 75% | Weight (kg) | EWH Nott, [37] | EWH Pepe et al. [38] |
|---|---|---|---|---|---|---|---|---|
| 1 | 32.0 | 20.0 | 16.0 | 10,240 | 7680 | 23.0 | 3.5 | 3.9 |
| 2 | 18.0 | 12.0 | 9.0 | 1944 | 1458 | 4.4 | 2.0 | 2.2 |
| 3 | 15.0 | 10.0 | 9.0 | 1350 | 1013 | 3.0 | 1.6 | 2.2 |
| 4 | 25.0 | 15.0 | 14.0 | 5250 | 3938 | 11.8 | 2.7 | 3.4 |
| 5 | 25.0 | 19.0 | 12.0 | 5700 | 4275 | 12.8 | 2.7 | 3.0 |
| 6 | 49.0 | 32.0 | 28.0 | 43,904 | 32,928 | 98.8 | 5.3 | 6.9 |
| 7 | 32.0 | 15.0 | 13.0 | 6240 | 4680 | 14.0 | 3.5 | 3.2 |
| 8 | 15.0 | 12.0 | 7.0 | 1260 | 945 | 2.8 | 1.6 | 1.7 |
| 9 | 11.0 | 6.0 | 5.0 | 330 | 248 | 0.7 | 1.2 | 1.2 |
| 10 | 11.0 | 11.0 | 4.0 | 484 | 363 | 1.1 | 1.2 | 1.0 |
| 11 | 16.0 | 9.0 | 8.0 | 1152 | 864 | 2.6 | 1.7 | 2.0 |
| 12 | 23.0 | 11.0 | 6.0 | 1518 | 1139 | 3.4 | 2.5 | 1.5 |
| 13 | 12.0 | 10.0 | 6.0 | 720 | 540 | 1.6 | 1.3 | 1.5 |
| 14 | 23.0 | 14.5 | 14.0 | 4669 | 3502 | 10.5 | 2.5 | 3.4 |
| 15 | 10.0 | 6.5 | 3.0 | 195 | 146 | 0.4 | 1.1 | 0.7 |
| 16 | 8.0 | 6.0 | 3.0 | 144 | 108 | 0.3 | 0.9 | 0.7 |
| 17 | 10.0 | 8.0 | 6.0 | 480 | 360 | 1.1 | 1.1 | 1.5 |
| 18 | 9.5 | 6.0 | 4.5 | 257 | 192 | 0.6 | 1.0 | 1.1 |
| 19 | 11.0 | 9.0 | 5.0 | 495 | 371 | 1.1 | 1.2 | 1.2 |
| 20 | 12.0 | 5.5 | 5.0 | 330 | 248 | 0.7 | 1.3 | 1.2 |
| 21 | 11.0 | 5.5 | 5.5 | 333 | 250 | 0.7 | 1.2 | 1.4 |
| 22 | 17.0 | 7.0 | 6.5 | 774 | 580 | 1.7 | 1.9 | 1.6 |
| 23 | 21.0 | 13.0 | 10.0 | 2730 | 2048 | 6.1 | 2.3 | 2.5 |
| 24 | 15.5 | 8.5 | 6.0 | 791 | 593 | 1.8 | 1.7 | 1.5 |
| 25 | 13.0 | 10.0 | 8.5 | 1105 | 829 | 2.5 | 1.4 | 2.1 |

**Table 6.** Quantification of boulder size, volume, and estimated weight from the Pleistocene (MIS 5e) conglomerate at site 6 at Ponta do Cedro, on the east shore of Santa Maria Island (see map, Figure 1b, and field photo, Figure 9). The standard density of basalt at 3.0 gm/cm$^3$ was applied uniformly in order to calculate wave height for each boulder. EWH: estimated wave height (in meters), calculated from equations in Nott [37] and Pepe et al. [38]. See the methods Section 3.2. (hydraulic model).

| Sample | Long Axis (cm) | Intermediate Axis (cm) | Short Axis (cm) | Volume (cm$^3$) | Adjust. to 75% | Weight (kg) | EWH Nott, [37] | EWH Pepe et al. [38] |
|---|---|---|---|---|---|---|---|---|
| 1 | 20.0 | 18.0 | 9.0 | 3240 | 2430 | 7.3 | 2.2 | 2.2 |
| 2 | 13.5 | 12.5 | 5.0 | 844 | 633 | 1.9 | 1.5 | 1.2 |
| 3 | 26.5 | 19.0 | 8.0 | 4028 | 3021 | 9.1 | 2.9 | 2.0 |
| 4 | 29.0 | 11.5 | 10.0 | 3335 | 2501 | 7.5 | 3.2 | 2.5 |
| 5 | 23.0 | 23.0 | 13.0 | 6877 | 5158 | 15.5 | 2.5 | 3.2 |
| 6 | 20.0 | 19.0 | 10.0 | 3800 | 2850 | 8.6 | 2.2 | 2.5 |
| 7 | 25.0 | 21.0 | 9.0 | 4725 | 3544 | 10.6 | 2.7 | 2.2 |
| 8 | 20.0 | 10.5 | 8.0 | 1680 | 1260 | 3.8 | 2.2 | 2.0 |
| 9 | 40.0 | 21.5 | 17.0 | 14,620 | 10,965 | 32.9 | 4.4 | 4.2 |
| 10 | 18.0 | 13.0 | 12.0 | 2808 | 2106 | 6.3 | 2.0 | 3.0 |
| 11 | 19.0 | 17.0 | 12.0 | 3876 | 2907 | 8.7 | 2.1 | 3.0 |
| 12 | 28.0 | 21.0 | 15.0 | 8820 | 6615 | 19.8 | 3.1 | 3.7 |
| 13 | 28.5 | 19.0 | 16.0 | 8664 | 6498 | 19.5 | 3.1 | 3.9 |
| 14 | 35.5 | 18.0 | 15.0 | 9585 | 7189 | 21.6 | 3.9 | 3.7 |
| 15 | 26.0 | 24.0 | 15.0 | 9360 | 7020 | 21.1 | 2.8 | 3.7 |
| 16 | 19.0 | 12.0 | 8.0 | 1824 | 1368 | 4.1 | 2.1 | 2.0 |
| 17 | 14.0 | 8.0 | 7.5 | 840 | 630 | 1.9 | 1.5 | 1.8 |
| 18 | 23.5 | 17.0 | 15.0 | 5993 | 4494 | 13.5 | 2.6 | 3.7 |
| 19 | 15.0 | 12.0 | 6.0 | 1080 | 810 | 2.4 | 1.6 | 1.5 |
| 20 | 30.0 | 26.0 | 26.0 | 20,280 | 15,210 | 45.6 | 3.3 | 6.4 |
| 21 | 27.0 | 15.0 | 15.0 | 6075 | 4556 | 13.7 | 2.9 | 3.7 |
| 22 | 20.0 | 12.0 | 10.0 | 2400 | 1800 | 5.4 | 2.2 | 2.5 |
| 23 | 17.0 | 12.0 | 8.0 | 1632 | 1224 | 3.7 | 1.9 | 2.0 |
| 24 | 31.0 | 17.0 | 11.0 | 5797 | 4348 | 13.0 | 3.4 | 2.7 |
| 25 | 56.0 | 36.0 | 16.0 | 32,256 | 24,192 | 72.6 | 6.1 | 3.9 |

**Table 7.** Quantification of boulder size, volume, and estimated weight from the Pleistocene conglomerate at site 7 at Ponta do Cedro, on the east shore of Santa Maria Island (see map, Figure 1b). The standard density of basalt at 3.0 gm/cm$^3$ was applied uniformly in order to calculate wave height for each boulder. EWH: estimated wave height (in meters), calculated from equations in Nott [37] and Pepe et al. [38]. See the methods Section 3.2. (hydraulic model).

| Sample | Long Axis (cm) | Intermediate Axis (cm) | Short Axis (cm) | Volume (cm$^3$) | Adjust. to 75% | Weight (kg) | EWH Nott, [37] | EWH Pepe et al. [38] |
|---|---|---|---|---|---|---|---|---|
| 1 | 50.0 | 40.0 | 25.0 | 50,000 | 37,500 | 112.5 | 5.5 | 6.2 |
| 2 | 28.0 | 24.5 | 20.0 | 13,720 | 10,290 | 30.9 | 3.1 | 4.9 |
| 3 | 50.0 | 36.0 | 22.0 | 39,600 | 29,700 | 89.1 | 5.5 | 5.4 |
| 4 | 27.0 | 21.0 | 10.0 | 5670 | 4253 | 12.8 | 2.9 | 2.5 |
| 5 | 38.0 | 21.0 | 20.0 | 15,960 | 11,970 | 35.9 | 4.1 | 4.9 |
| 6 | 26.0 | 18.5 | 13.0 | 6253 | 4690 | 14.1 | 2.8 | 3.2 |
| 7 | 17.0 | 15.0 | 11.0 | 2805 | 2104 | 6.3 | 1.9 | 2.7 |
| 8 | 27.0 | 23.0 | 13.5 | 8384 | 6288 | 18.9 | 2.9 | 3.3 |
| 9 | 38.0 | 32.0 | 23.0 | 27,968 | 20,976 | 62.9 | 4.1 | 5.7 |
| 10 | 20.0 | 16.0 | 7.5 | 2400 | 1800 | 5.4 | 2.2 | 1.8 |
| 11 | 41.0 | 26.0 | 25.0 | 26,650 | 19,988 | 60.0 | 4.5 | 6.2 |
| 12 | 27.0 | 22.0 | 20.0 | 11,880 | 8910 | 26.7 | 2.9 | 4.9 |
| 13 | 26.0 | 20.0 | 15.0 | 7800 | 5850 | 17.6 | 2.8 | 3.7 |
| 14 | 25.0 | 16.0 | 9.0 | 3600 | 2700 | 8.1 | 2.7 | 2.2 |
| 15 | 20.0 | 15.0 | 9.0 | 2700 | 2025 | 6.1 | 2.2 | 2.2 |
| 16 | 19.0 | 16.0 | 10.0 | 3040 | 2280 | 6.8 | 2.1 | 2.5 |
| 17 | 15.0 | 10.5 | 10.0 | 1575 | 1181 | 3.5 | 1.6 | 2.5 |
| 18 | 16.5 | 13.0 | 6.0 | 1287 | 965 | 2.9 | 1.8 | 1.5 |
| 19 | 16.0 | 11.5 | 3.0 | 552 | 414 | 1.2 | 1.7 | 0.7 |
| 20 | 14.0 | 9.0 | 7.5 | 945 | 709 | 2.1 | 1.5 | 1.8 |
| 21 | 15.5 | 11.0 | 8.0 | 1364 | 1023 | 3.1 | 1.7 | 2.0 |
| 22 | 21.0 | 17.5 | 15.5 | 5696 | 4272 | 12.8 | 2.3 | 3.8 |
| 23 | 29.0 | 16.0 | 9.0 | 4176 | 3132 | 9.4 | 3.2 | 2.2 |
| 24 | 22.5 | 14.0 | 12.0 | 3780 | 2835 | 8.5 | 2.5 | 3.0 |
| 25 | 30.0 | 21.0 | 13.0 | 8190 | 6143 | 18.4 | 3.3 | 3.2 |

Data points representing individual boulders sampled from the conglomerate layer at Ponta do Castelo, correlated at sites 5 to 7, were plotted on a set of Sneed-Folk triangular diagrams (Figure 4e–g).

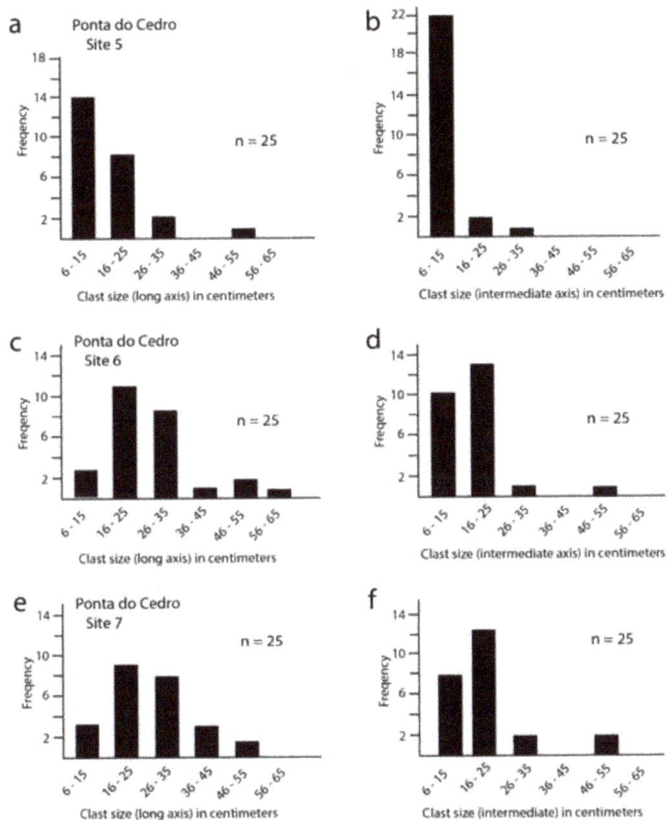

**Figure 9.** Bar graphs used to appraise variations in the long and intermediate axes on basalt clasts from three sites of correlated Pleistocene CBDs at Ponta do Cedro; (**a**) Long-axis length from clasts at site 5; (**b**) Intermediate-axis length from clasts at site 5; (**c**) Long-axis length from clasts at site 6; (**d**) Intermediate-axis length from clasts at site 6; (**e**) Long-axis length from clasts at site 7; (**f**) Intermediate-axis length from clasts at site 7.

Overall, the distribution of clasts among the three Ponta do Cedro samples was highly consistent, with 68–78% of data points limited to the two central blocks within the triangular plot. The occurrence of data points in the top triangle, as well as the lower-right corner of the plot, were equally rare. Even so, there was a tendency for the data clusters to slope towards the right, indicative of a slight favorability towards elongated shapes. This trend was not nearly as strong as detected in modern CBDs at Prainha (Figure 4a) or Ponta do Castelo (Figure 4c), but entirely consistent with trends in the Pleistocene (MIS 5e) CBDs at Prainha (Figure 4b) or Ponta do Castelo (Figure 4d).

Variations in boulder size as a function of maximum and intermediate axis length were plotted using bar graphs (Figure 9), based on raw data drawn from Tables 5–7. The percentage of boulders in each Pleistocene sample was found to increase from the 20% (the more southern locality at site 5) to 48% (at site 6) up to the maximum value of 56% (the more northern locality at site 7). In each case, there was a marked shift in the abundance of clasts measured on the intermediate axis allocated to the size range between 6 and 15 cm (Figure 9b,d,f). Such a trend reinforced the impression that the Pleistocene (MIS 5e) basalt clasts conformed to a moderately oblong shape. Site 5 (Figure 9a,b) stood out among all the Pleistocene samples as most dominated by clasts defined by cobble size. In contrast, sites 6 and 7 were remarkably similar in their size distributions. Moreover, these samples were indistinguishable from the Pleistocene CBD sample at Ponta do Castelo, which also included a high concentration of boulders.

### 4.6. Context of the Pleistocene Fauna at Ponta Do Cedro

At Ponta do Cedro, preliminary work allowed us to report a total of 13 Last Interglacial mollusk taxa (12 gastropods and 1 bivalve species). As in other MIS 5e outcrops (e.g., Prainha, Vinha Velha, Lagoinhas), the thermophilic element was present through several species of *Conus* and of the Pisaniidae *Gemophos viverratus* (Kiener, 1834) (=*Cantharus variegatus*). Ponta do Cedro was classified as a fossiliferous geosite of regional relevance [15,16], a situation that would change in the near future, as a result of ongoing work.

### 4.7. Analysis of Calculated Storm-Wave Heights

A summary of key data was provided (Table 8), pertaining to average boulder size and maximum boulder size from the five Pleistocene (MIS 5e) transects and two modern transects as correlated with weight calculated on the basis of specific gravity for basalt, using the equations derived by Nott [37] and Pepe et al. [38]. These data were applied to estimate the wave heights required to transport boulders from the bedrock source in sea cliffs to their resting place, and variations in the results depended on the equations used.

**Table 8.** Summary data from Tables 1–7, showing maximum boulder size and estimated weight compared to the average values for all boulders (n = 25) from each of transects 1–7 together with calculated values for wave heights estimated as necessary for boulder mobility. Estimated average wave height (EAWH; in meters) and estimated maximum wave height (EMWH; in meters), calculated according to equations derived by Nott [37] and Pepe et al. [38]. (see the methods Section 3.2. (hydraulic model)). Maximum values for each Transect in bold.

| Transect | Locality | Age of the CBD | Average Boulder Size (cm$^3$) | Average Boulder Weight (kg) | Max Boulder Weight (kg) | EAWH Nott, [37] | EAWH Pepe et al. [38] | EMWH Nott, [37] | EMWH Pepe et al. [38] |
|---|---|---|---|---|---|---|---|---|---|
| 1 | Prainha | Modern | 7490 | 22.5 | 67.9 | **3.6** | 2.9 | 5.2 | 6.4 |
| 2 | Prainha | MIS 5e | 476 | 1.4 | 9.9 | 1.3 | 1.3 | 2.7 | 3.2 |
| 3 | Ponta do Castelo | Modern | 2713 | 8.1 | 18.6 | 2.8 | 2.0 | 5.5 | 3.7 |
| 4 | Ponta do Castelo | MIS 5e | 6731 | 20.2 | 101.4 | 2.7 | **3.3** | 5.3 | **8.1** |
| 5 | Ponta do Cedro | MIS 5e | 2772 | 8.3 | 98.8 | 1.9 | 2.1 | 5.3 | 6.9 |
| 6 | Ponta do Cedro | MIS 5e | 4933 | 14.8 | 72.6 | 2.7 | 2.9 | **6.1** | 6.4 |
| 7 | Ponta do Cedro | MIS 5e | **7680** | **23.0** | **112.5** | 2.9 | 3.3 | 5.5 | 6.2 |

First, we have discussed the estimates using Nott's equation. The estimated wave height needed to move the largest boulder encountered in the Pleistocene (MIS 5e) transects (transect 6 at Ponta do Cedro) amounted to 6.1 m. The comparison between the Prainha and Ponta do Castelo modern and the Pleistocene (MIS 5e) CBD gave contrasting results; at Prainha, the average boulder size and weight and estimated average wave height were higher in the modern CBD in comparison with the Pleistocene CBD, whereas at Ponta do Castelo, it was the opposite trend. Moreover, the estimated maximum wave height required to shift boulders was very similar in all sites (ranging from 5.2 to 6.1 m) but Prainha (MIS 5e), where this value was about half (2.7 m). There was also a large difference between the largest boulders from Prainha (MIS 5e) and Ponta do Castelo (modern) CBDs, which were much smaller than those from the other sites (Table 8).

The results using the equation of Pepe et al. [38] were different from those using Nott's equation [37], as the former consistently indicated that the maximum values for both the estimated average wave height (EAWH) and the estimated maximum wave height (EMWH) occurred at the MIS 5e transects (at Ponta do Castelo and Ponta do Cedro, in the case of the EAWH; at Ponta do Castelo, in the case of the EMWH; cf. Table 8).

## 5. Discussion

### 5.1. CBD: Tsunami Versus Storms, and the Use of Flawed Equations

A recently published field-test of the hydrodynamic equations, based on the Nott-Approach and their derivations [51–54], failed to validate estimations of the wave height from boulder dimensions and concluded that such equations were flawed because they yielded unrealistically large heights [55]. These authors also stated that the Nott-Approach analysis did not differentiate between storm and tsunami waves [55]. Such criticism might be valid, but acceptance in full threatens to nullify any attempt to engage in the quantification of CBDs from the geologic record. The Nott-Approach remains useful and is applied solely to compare the relative storminess imprint left in CBDs from different time-periods (MIS 5e vs. modern, as in our example), but located at the same site. For this particular case and question, it does not matter that equations may be flawed because the errors are uniformly carried through the exercise. What is important is that such equations function to detect consistent differences between the estimated wave heights for the past (MIS 5e) and for the modern situation. Moreover, they also provide the magnitude of possible differences.

CBDs should be portrayed as archives of extreme wave events, and their origin may be due to either storm waves or tsunamis. All CBDs studied in Santa Maria Island have no characteristics of tsunami deposits (e.g., imbrication of the boulders, erosive base (scour-and-fill features), rip-up clasts of the underlying substratum, downward-injected clastic dykes inside the palaeosol, traction figures; [56]), and we have, therefore, interpreted them as the result of storm waves.

The width of Santa Maria Island's insular platform is larger in the N and W (although this is not apparent in the western shores, as a result of the uplift of the island) due to the higher offshore significant wave heights mainly coming from the NW, W, and N [24]; in contrast, at the south and eastern shores, the insular platform is quite narrow [25]. However, the most destructive storms (and, therefore, with higher wave heights) that impacted the Azores in recent times (and we assume a similar pattern in the past) were related to the passage of hurricanes. As these have a counter-clockwise rotation in the northern hemisphere, its passage most strongly affects precisely the southern and eastern shores of the islands, thus providing a possible explanation for the limited occurrence of CBD. Based on the historical record, recent hurricanes that struck the islands in the Azores Archipelago include "Hurricane 15" (1932), Hannah (1959), Debbie (1961), Fran (1973), Gordon (2006 and 2012), Gaston (2016), Ophelia (2017), and Lorenzo (2019). A direct hit on Santa Maria Island at 37° N entails a northward shift in the latitude of 10° and a longitudinal travel distance of roughly 2000 km. Hurricane frequency in the Azores is regarded as on the rise during the last 50 years [30]. The strong Pliocene sedimentary record on Santa Maria reflects some indication of hurricane activity on the southern coast [57], where the marine shelf is narrowest (Figure 1b). Moreover, the island's geological record holds evidences of Pliocene deep wave interaction with the seafloor, reaching 50 m depth (Ponta do Castelo, [21]), thus suggesting a high likelihood of large-size boulder transport.

During the Last Interglacial, the average values for summer insolation were about 11% higher than those of today [58], the oceanic Polar Front was pushed north-westwards by the Gulf Stream warmer sea surface temperatures (SSTs) [59], and the summer position of the Azores High was forced E of its present location to around 35°N and 20–25°W, i.e., between the Azores and Iberia [60]. These oceanographic conditions induced a stronger North Atlantic storm track, shifted northward from its present location, and extended to the east [4]. Altogether, the storminess associated with tropical cyclones is expected to have been higher during the MIS 5e than today [1,3,61], as a result of the North Atlantic Subtropical High eastward shifting from its present location. However, the occurrence of "superstorms" during the Last Interglacial has been recently questioned by authors [62,63]. Therefore, probably, the higher storminess that most authors suggest for the Last Interglacial is related to a higher frequency of hurricanes and not necessarily with higher wave heights.

*5.2. Comparison between Model Wave-Height Data and Inferred Modern Wave-Height*

Our study yielded modern CBD estimated averages for wave heights varying from 2.8 to 3.6 m (using Nott's Equation (1)) and from 2.0 to 2.9 (Pepe et al.'s Equation (2)) (Table 8). The former values were higher than the mean values presented by Rusu and Onea [31] for the archipelago, which varied from 2.23 to 2.55 m, a situation that other authors also reported [55], with Nott's Equation consistently yielding wave height values higher than those registered by modern wave buoy. In contrast, the maximum calculated estimate wave heights of 5.5 m (Equation (1)) and 6.4 m (Equation (2)) (Table 8) did not get close to the lowest maximum value of Rusu and Onea [31] of 9.18 m.

The values obtained for the Pleistocene (MIS 5e) estimates were similar using both formulas, with wave height estimates varying from 1.3 to 2.9 m (Equation (1)) and from 1.3 to 3.3 m (Equation (2)); however, the maximum wave height estimates were quite different: 6.1 m (with Equation (1)) and 8.1 m (Equation (2)) (Table 8). These discrepancies highlighted the serious reserves [55], regarding the use of these formulas when solely targeted to the inference of modern (or past) wave heights, based on the dimensions/mass of the boulders.

Considering the general morphology of the sites, the nature of the boulders, and the processes that lead to its shaping, and albeit based on only two sites (Prainha and Ponta do Castelo; cf. Table 8), where it was possible to compare the modern CBD with the MIS 5e CBD, the results showed contrasting site-dependency. At Prainha, modern storminess was estimated to be higher during the Last Interglacial (independent of which equation was used), whereas at Ponta do Castelo, MIS 5e storminess should have been highly similar to the modern one (using Equation (1)) or higher (Equation (2)).

Further analysis is needed from other islands having similar MIS 5e/modern sites situation before sound conclusions might be reached regarding such a restrictive use of formulas, otherwise argued to be flawed [55]. Finally, a set of parallel profiles to the sea (at different heights) should be undertaken with respect to modern CBDs, as well as perpendicular profiles, in order to check for possible boulder size changes according to distance from present-day sea level. This analysis could also help to relate a mean boulder dimension to an average distance to the sea. An in-depth analysis of Pleistocene (MIS 5e) wave-cut surfaces will also enhance our knowledge of coastal retreat, average erosion rates, and wave activity.

*5.3. Comparison with CBD Studies Elsewhere*

A recent review [64] indicated that although significant studies have dealt with the Quaternary CBDs, only 21 papers studied the Neogene deposits (10 of Miocene and 9 of Pliocene age, plus 2 works dealing with both epochs), mostly due to their low preservation potential [64]. According to our literature survey, this was the first study on Pleistocene CBDs conducted on the Atlantic oceanic islands. Moreover, previous works using the same methodology dealt with the different rocky composition of the CBDs (limestone, rhyolite, and andesite) [17–19]; this paper being also the first to use the same kind of analyses in regard to basalt, which has higher specific gravity than any of the other rocks tested. Therefore, for similar size CBDs, basalt boulders will require a more energetic wave to be moved.

## 6. Conclusions

Limited to comparative results from modern and late Pleistocene (MIS 5e) boulder deposits on the southern and eastern shores of Santa Marina Island, the present study permitted the following conclusions:

1) Wave heights estimated on the basis of the largest modern boulders from CBDs on the modern south shore of Santa Maria Island indicated maximum values between 5.2 and 5.5 m (Equation (1)) and between 3.7 and 6.4 m (Equation (2)), which was 2 m higher than the expected average height experienced during winter storms, supporting the conclusions of [55] regarding the advice on the non-use of these formulas because of unrealistic, flawed estimates of wave heights;

2) Regarding storminess, our results were contradictory and did not allow any conclusion about possible higher storminess during the Last Interglacial when compared with the present-day, and recorded in the CBDs of Santa Maria Island;

3) Historical records of storm activity, coupled with the westerlies regime predominant in the Azores area, placed a stronger emphasis on the seasonal effect of winter storms that preferentially influence shore erosion on north-facing coasts. Although the historical record indicates that hurricane activity is less frequent in the islands, its erosional effect against a south- and eastern-facing shores must be considered. From the available data derived from CBDs and coastal geomorphology around Santa Maria Island, it was clear that wave action both today and approximately 125,000 years ago during MIS 5e remained consistently highest against the eastern and south-eastern coasts at Ponta do Cedro and Ponta do Castelo. These localities also correspond to areas where the island's marine shelf is narrowest. Although the historical incidence of hurricanes passing through the Azores Archipelago is statistically low compared to the arrival of annual winter storms, the migration of hurricanes moving in a counter-clockwise rotation across the North Atlantic Ocean conforms to the evidence for excessive erosion rates in those districts beyond the capacity of winter storms.

The abundant fossil record on Santa Marina Island that makes accurate dating possible for the Late Pleistocene (MIS 5e) interglacial epoch is not available elsewhere in the Azores Archipelago, but future research on MIS 5e versus modern CBDs (whenever located in the same site) from other archipelagos is expected to apply the same formulaic techniques to extract information on storminess.

**Author Contributions:** Fieldwork in Santa Maria Island has occurred since 1999 by a large, multidisciplinary team led by S.P.Á. during the workshops "Palaeontology in Atlantic Islands". Field data on CBDs was collected in July 2019 by A.C.R., L.B., and C.S.M., M.E.J. prepared the first draft of this contribution and drafted all figures, except for Figure 1 prepared by C.S.M., S.P.Á. was responsible for working out the mathematics related to storm hydrodynamics, produced all tables, and supplied all ground photos. S.P.Á. and C.S.M. summarized the literature on CBDs in the Azores region. In addition, C.S.M. contributed input on wave regimes and main directions within the study area. Authorship has been limited to those who contributed substantially to the work reported. All authors have read and agreed to the published version of the manuscript.

**Funding:** We thank Direcção Regional da Ciência e Tecnologia (Regional Government of the Azores), FCT (Fundação para a Ciência e a Tecnologia) of the Portuguese Government, and Câmara Municipal de Vila do Porto for financial support. S.P.Á. acknowledges his research contract (IF/00465/2015) funded by the Portuguese Science Foundation (FCT). A.C.R. was supported by a Post-doctoral grant SFRH/BPD/117810/2016 from FCT. L.B. acknowledges her Ph.D. Grant from FCT (SFRH/BD/135918/2018). C.S.M. acknowledges his Ph.D. Grant M3.1.a/F/100/2015 from Fundo Regional para a Ciência e Tecnologia (FRCT). This work was also supported by FEDER funds through the Operational Programme for Competitiveness Factors – COMPETE and by National Funds through FCT under the UID/BIA/50027/2013, POCI-01-0145-FEDER-006821, and under DRCT-M1.1.a/005/Funcionamento-C-/2016 (CIBIO-A) project from FRCT. This work was also supported by FEDER funds (in 85%) and by funds from the Regional Government of the Azores (15%) through Programa Operacional Açores 2020 under the project VRPROTO (ACORES-01-0145-FEDER-000078).

**Acknowledgments:** We acknowledge the field assistance of Manta Maria and Câmara Municipal de Vila do Porto. We are grateful to the organizers and participants of all editions of the international workshops "Palaeontology in Atlantic Islands", who helped with fieldwork (2002–2019). We are grateful to Daniele Casalbore and two anonymous reviewers, whose comments greatly improved this manuscript.

**Conflicts of Interest:** The authors declare no conflict of interest.

# References

1. Hansen, J.; Sato, M.; Hearty, P.; Ruedy, R.; Kelley, M.; Masson-Delmotte, V.; Russell, G.; Tselioudis, G.; Cao, J.; Rignot, E.; et al. Ice melt, sea level rise and superstorms: Evidence from paleoclimate data, climate modeling, and modern observations that 2 °C global warming could be dangerous. *Atmos. Chem. Phys.* **2016**, *16*, 3761–3812. [CrossRef]

2. Hearty, P.J.; Neumann, A.C. Rapid sea level and climate change at the close of the Last Interglaciation (MIS 5e): Evidence from the Bahama Islands. *Quat. Sci. Rev.* **2001**, *20*, 1881–1895. [CrossRef]

3. Hearty, P.J.; Tormey, B.R. Sea-level change and super storms; geologic evidence from the last interglacial (MIS 5e) in the Bahamas and Bermuda offers ominous prospects for a warming Earth. *Mar. Geol.* **2017**, *390*, 347–365. [CrossRef]

4.  Ávila, S.P.; Melo, C.; Silva, L.; Ramalho, R.S.; Quatau, R.; Hipólito, A.; Cordeiro, R.; Rebelo, A.C.; Madeira, P.; Rovere, A.; et al. A review of the MIS 5e highstand deposits from Santa Maria Island (Azores, NE Atlantic): Palaeobiodiversity, palaeoecology and palaeobiogeography. *Quat. Sci. Rev.* **2015**, *114*, 126–148. [CrossRef]

5.  Naylor, L.A.; Stephenson, W.J. On the role of discontinuities in mediating shore platform erosion. *Geomorphology* **2011**, *114*, 89–100. [CrossRef]

6.  Borges, P. Ambientes Litorais nos Grupos Central e Oriental do Arquipélago dos Açores, Conteúdos e Dinâmica de Microescala. Ph.D. Thesis, Universidade dos Açores, Ponta Delgada, Portugal, 2003; pp. 1–412.

7.  Quartau, R.; Trenhaile, A.S.; Mitchell, N.C.; Tempera, F. Development of volcanic insular shelves: Insights from observations and modelling of Faial Island in the Azores Archipelago. *Mar. Geol.* **2010**, *275*, 66–83. [CrossRef]

8.  Ramalho, R.S.; Helffrich, G.; Madeira, J.; Cosca, M.; Thoas, C.; Quartau, R.; Hipólito, A.; Rovere, A.; Hearty, P.J.; Ávila, S.P. Emergence and evolution of Santa Maria Island (Azores)—The conundrum of uplifted islands revisited. *Geol. Soc. Am. Bull.* **2017**, *129*, 372–391. [CrossRef]

9.  Casalbore, D.; Romagnoli, C.; Bosman, A.; Anzidei, M.; Chiocci, F.L. Coastal hazard due to submarine canyons in active insular volcanoes: Examples from Lipari Island (southern Tyrrhenian Sea). *J. Coast. Conserv.* **2018**, *22*, 989–999. [CrossRef]

10. Bosman, A.; Casalbore, D.; Romagnoli, C.; Chiocci, F.L. Formation of an 'a'ā lava delta: Insights from time-lapse multibeam bathymetry and direct observations during the Stromboli 2007 eruption. *Bull. Volc.* **2014**, *76*, 838. [CrossRef]

11. Rusu, L.; Soares, C.G. Wave energy assessments in the Azores islands. *Renew. Energy* **2012**, *45*, 183–196. [CrossRef]

12. Ávila, S.P.; Melo, C.; Berning, B.; Sá, N.; Quartau, R.; Rijsdijk, K.F.; Ramalho, R.S.; Cordeiro, R.; De Sá, N.C.; Pimentel, A.; et al. Towards a 'Sea-Level Sensitive' dynamic model: Impact of island ontogeny and glacio-eustasy on global patterns of marine island biogeography. *Biol. Rev.* **2019**, *94*, 1116–1142.

13. Melo, C.S.; Ramalho, R.S.; Quartau, R.; Hipólito, A.; Gil, A.; Borges, P.A.; Cardigos, F.; Ávila, S.P.; Madeira, J.; Gaspar, J.L. Genesis and morphological evolution of coastal talus-platforms (fajãs) with lagoons: The case study of the newly-formed Fajã dos Milagres (Corvo Island, Azores). *Geomorphology* **2018**, *310*, 138–152. [CrossRef]

14. Madeira, P.; Kroh, A.; Cordeiro, R.; Meireles, R.; Ávila, S.P. The fossil echinoids of Santa Maria island, Azores (Northern Atlantic Ocean). *Acta Geol Pol* **2011**, *61*, 243–264.

15. Ávila, S.P.; Cachão, M.; Ramalho, R.S.; Botelho, A.Z.; Madeira, P.; Rebelo, A.C.; Cordeiro, R.; Melo, C.; Hipólito, A.; Ventura, M.A.; et al. The Palaeontological heritage of Santa Maria Island (Azores: NE Atlantic): A re-evaluation of geosites in GeoPark Azores and their use in geotourism. *Geoheritage* **2016**, *8*, 155–171. [CrossRef]

16. Raposo, V.; Melo, C.; Ventura, M.A.; Câmara, R.; Tavares, J.; Johnson, M.E.; Ávila, S.P. Comparing methods of evaluation of geosites: The fossiliferous outcrops of Santa Maria Island (Azores: NE Atlantic) as a case-study for sustainable Island tourism. *Sustainability* **2018**, *10*, 3596. [CrossRef]

17. Johnson, M.E.; Ledesma-Vázquez, J.; Guardado-France, R. Coastal Geomorphology of a Holocene Hurricane Deposit on a Pleistocene Marine Terrace from Isla Carmen (Baja California Sur, Mexico). *J. Mar. Sci. Eng.* **2018**, *6*, 108. [CrossRef]

18. Johnson, M.E.; Guardado-France, R.; Johnson, E.M.; Ledesma-Vázquez, J. Geomorphology of a Holocene Hurricane Deposit Eroded from Rhyolite Sea Cliffs on Ensenada Almeja (Baja California Sur, Mexico). *J. Mar. Sci. Eng.* **2019**, *7*, 193. [CrossRef]

19. Johnson, E.M.; Johnson, E.M.; Guardado-France, R.; Ledesma-Vázquez, J. Holocene Hurricane Deposits Eroded as Coastal Barriers from Andesite Sea Cliffs at Puerto Escondido (Baja California Sur, Mexico). *J. Mar. Sci. Eng.* **2020**, *8*, 75. [CrossRef]

20. Vogt, P.R.; Jung, W.-P. The "Azores Geosyndrome" and plate tectonics: Research history, synthesis, and unsolved puzzles. In *Volcanoes of the Azores*; Kueppers, U., Beier, C., Eds.; Active Volcanoes of the World Series; Springer: Berlin/Heidelberg, Germany, 2018; pp. 27–56.

21. Meireles, R.P.; Quartau, R.; Ramalho, R.; Madeira, J.; Rebelo, A.C.; Zanon, V.; Ávila, S.P. Depositional processes on oceanic island shelves—Evidence from storm-generated Neogene deposits from the mid-North Atlantic. *Sedimentology* **2013**, *60*, 1769–1785. [CrossRef]

22. Ávila, S.P.; Ramalho, R.; Habermann, J.; Quartau, R.; Kroh, A.; Berning, B.; Johnson, M.; Kirby, M.; Zanon, V.; Titschack, J.; et al. Palaeoecology, taphonomy, and preservation of a lower Pliocene shell bed (coquina) from a volcanic oceanic island (Santa Maria Island, Azores, NE Atlantic Ocean). *Palaeogeogr. Palaeoclimatol. Palaeoecol.* **2015**, *430*, 57–73. [CrossRef]

23. Ávila, S.P.; Ramalho, R.; Habermann, J.; Titschack, J. The marine fossil record at Santa Maria Island (Azores). In *Volcanoes of the Azores*; Kueppers, U., Beier, C., Eds.; Active Volcanoes of the World Series; Springer: Berlin/Heidelberg, Germany, 2018; pp. 155–196.

24. Ricchi, A.; Quartau, R.; Ramalho, R.S.; Romagnoli, C.; Casalbore, D.; Zhao, Z. Imprints of volcanic, erosional, depositional, tectonic and mass-wasting processes in the morphology of Santa Maria insular shelf. *Mar. Geol.* **2020**, *424*, 106163. [CrossRef]

25. Ricchi, A.; Quartau, R.; Ramalho, R.S.; Romagnoli, C.; Casalbore, D.; da Cruz, J.V.; Fradique, C.R.; Vinhas, A. Marine terrace development on reefless volcanic islands: New insights from high-resolution marine geophysical data offshore Santa Maria Island (Azores Archipelago). *Mar. Geol.* **2018**, *406*, 42–56. [CrossRef]

26. Lourenço, N.; Miranda, J.M.; Luis, J.F.; Ribeiro, A.; Mendes Victor, L.A.; Madeira, J.; Needham, H.D. Morpho-tectonic analysis of the Azores volcanic plateau from a new bathymetric compilation of the area. *Mar. Geophys. Res.* **1998**, *20*, 141–156. [CrossRef]

27. Madeira, J.; Ribeiro, A. Geodynamic models for the Azores triple junction: A contribution from tectonics. *Tectonophysics* **1990**, *184*, 405–415. [CrossRef]

28. Madeira, J. Estudos de Neotectónica nas Ilhas do Faial, Pico e S. Jorge: Uma Contribuição para o Conhecimento Geodinâmico da Junção Tripla dos Açores. Ph.D. Thesis, Lisbon University, Lisbon, Portugal, 1998; pp. 1–481.

29. Brito de Azevedo, E. Condicionantes dinâmicas do clima do Arquipélago dos Açores: Elementos para o seu estudo. *Açoreana* **2001**, *9*, 309–317.

30. Andrade, C.; Trigo, R.M.; Freitas, M.C.; Callego, M.C.; Borges, P.; Ramos, A.M. Comparing historic records of storm frequency and the North Atlantic Ocillation (NAO) chronology for the Azores region. *Holocene* **2008**, *18*, 745–754. [CrossRef]

31. Rusu, E.; Onea, F. Estimation of the wave energy conversion efficiency in the Atlantic Ocean close to the European islands. *Renew. Energy* **2016**, *85*, 687–703. [CrossRef]

32. Gould, W.J. Physical oceanography of the Azores front. *Prog. Oceanogr.* **1985**, *14*, 167. [CrossRef]

33. Klein, B.; Siedler, G. On the origin of the Azores Current. *J. Geophys. Res.* **1989**, *94*, 6159–6168. [CrossRef]

34. Quartau, R. The Insular Shelf of Faial: Morphological and Sedimentary Evolution. Ph.D. Thesis, Universidade de Aveiro, Aveiro, Portugal, 2007; pp. 1–326.

35. Wentworth, C.K. A scale of grade and class terms for clastic sediments. *J. Geol.* **1922**, *27*, 377–392. [CrossRef]

36. Sneed, E.D.; Folk, R.L. Pebbles in the lower Colorado River of Texas: A study in particle morphogenesis. *J. Mar. Geol.* **1958**, *66*, 114–150. [CrossRef]

37. Nott, J. Waves, coastal bolder deposits and the importance of pre-transport setting. *Earth Planet. Sci. Lett.* **2003**, *210*, 269–276. [CrossRef]

38. Pepe, F.; Corradino, M.; Parrino, N.; Besio, G.; Presti, V.L.; Renda, P.; Calcagnile, L.; Quarta, G.; Sulli, A.; Antonioli, F. Boulder coastal deposits at Favignana Island rocky coast (Sicily, Italy): Litho-structural and hydrodynamic control. *Geomorphology* **2018**, *303*, 191–209. [CrossRef]

39. Nandasena, N.A.K.; Paris, R.; Tanaka, N. Reassessment of hydrodynamic equations: Minimum flow velocity to initiate boulder transport by high energy events (storms, tsunamis). *Mar. Geol.* **2011**, *281*, 70–84. [CrossRef]

40. Ávila, S.P.; Rebelo, A.; Medeiros, A.; Melo, C.; Gomes, C.; Bagaço, L.; Madeira, P.; Borges, P.A.; Monteiro, P.; Cordeiro, R.; et al. *Os Fósseis de Santa Maria (Açores). 1. A Jazida da Prainha*; OVGA e Observatório Vulcanológico e Geotérmico dos Açores: Lagoa, Portugal, 2010; pp. 1–103.

41. Zbyszewsky, G.; Ferreira, O.; da, V. La faune marine des basses plages quaternaires de Praia et Prainha dans l'ile de Santa Maria (Açores). *Comunicações dos Serviços Geológicos de Portugal* **1961**, *45*, 467–478.

42. Zbyszewsky, G.; Ferreira, O.; da, V. Étude géologique de l'île de Santa Maria (Açores). *Comunicações dos Serviços Geológicos de Portugal* **1962**, *46*, 209–245.

43. García-Talavera, F. Fauna tropical en el Neotirreniense de Santa Maria (I. Azores). *Lavori S.I.M.* **1990**, *23*, 439–443.

44. Ávila, S.P.; Amen, R.; Azevedo, J.M.N.; Cachão, M.; García-Talavera, F. Checklist of the Pleistocene marine molluscs of Prainha and Lagoinhas (Santa Maria Island, Azores). *Açoreana* **2002**, *9*, 343–370.

45. Ávila, S.P.; Madeira, P.; Mendes, N.; Rebelo, A.; Medeiros, A.; Gomes, C.; García-Talavera, F.; Marques da Silva, C.; Cachão, M.; Martins, A.M.F. Luria lurida (Mollusca: Gastropoda), a new record for the Pleistocene of Santa Maria (Azores, Portugal). *Arquipélago* **2007**, *24*, 53–56.

46. Ávila, S.P.; Madeira, P.; Zazo, C.; Zazo Kroh, A.; Kirby, M.; Silva, C.M.; da Cachão, M.; Martins, A.M.F. Palaeocology of the Pleistocene (MIS 5.5) outcrops of Santa Maria Island (Azores) in a complex oceanic tectonic setting. *Palaeogeogr. Palaeoclimatol. Palaeoecol.* **2009**, *274*, 18–31. [CrossRef]

47. Ávila, S.P.; Silva CM da Schiebel, R.; Cecca, F.; Backeljau, T.; Martins, A.M.F. How did they get here? Palaeobiogeography of the Pleistocene marine molluscs of the Azores. *Bull. Geol. Soc. Fr.* **2009**, *180*, 295–307. [CrossRef]

48. Amen, R.G.; Neto, A.I.; Azevedo, J.M.N. Coralline-algal framework in the Quaternary of Prainha (Santa Maria Island, Azores). *Rev. Española Micropaleontol.* **2005**, *37*, 63–70.

49. Ávila, S.P.; Azevedo, J.M.N.; Madeira, P.; Cordeiro, R.; Melo, C.S.; Baptista, L.; Torres, P.; Johnson, M.E.; Vullo, R. Pliocene and Late Pleistocene actinopterygian fishes from Santa Maria Island, Azores (NE Atlantic Ocean): Palaeoecological and palaeobiogeographical implications. *Geol. Mag.* **2020**. [CrossRef]

50. Ávila, S.P.; Cordeiro, R.; Rodrigues, R.; Rebelo, A.C.; Melo, C.; Madeira, P.; Pyenson, N.D. Fossil Mysticeti from the Pleistocene of Santa Maria Island, Azores (NE Atlantic Ocean), and the prevalence of fossil cetaceans on oceanic islands. *Palaeontol. Electron.* **2015**, *18.2.27A*, 1–12.

51. Barbano, M.S.; Pirrotta, C.; Gerardi, F. Large boulders along the south-eastern Ionian coast of Sicily: Storm or tsunami deposits? *Mar. Geol.* **2010**, *275*, 140–154. [CrossRef]

52. Benner, R.; Browne, T.; Brückner, H.; Kelletat, D.; Scheffers, A. Boulder transport by waves: Progress in physical modelling. *Z. Geomorphol.* **2010**, *4* (Suppl. 3), 127–146. [CrossRef]

53. Deguara, J.C.; Gauci, R. Evidence of extreme wave events from boulder deposits on the south-east coast of Malta (Central Mediterranean). *Nat. Hazards* **2017**, *86*, 543–568. [CrossRef]

54. Engel, M.; May, S.M. Bonaire's boulder fields revisited: Evidence for Holocene tsunami impact on the Leeward Antilles. *Quat. Sci. Rev.* **2012**, *54*, 126–141. [CrossRef]

55. Cox, R.; Ardhuin, F.; Dias, F.; Autret, R.; Beisiegel, N.; Earlie, C.S.; Herterich, J.G.; Kennedy, A.; Paris, R.; Raby, A.; et al. Systematic review shows that work done by storm waves can be misinterpreted as tsunami-related because commonly used hydrodynamic equations are flawed. *Front. Mar. Sci.* **2020**, *7*, 4. [CrossRef]

56. Paris, R.; Ramalho, R.S.; Madeira, J.; Ávila, S.P.; May, S.M.; Rixhon, G.; Engel, M.; Brückner, H.; Herzog, M.; Schukraft, G.; et al. Mega-tsunami conglomerates and flank collapses of ocean island volcanoes. *Mar. Geol.* **2018**, *395*, 168–187. [CrossRef]

57. Johnson, M.E.; Uchman, A.; Costa, P.J.M.; Ramalho, R.S.; Ávila, S.P. Intense hurricane transports sand onshore: Example from the Plioclene Malbusca section on Stanta Maria Island (Azores, Portugal). *Mar. Geol.* **2017**, *385*, 244–249. [CrossRef]

58. CAPE. Last Interglacial, Project Members. Last Interglacial Arctic warmth confirms polar amplification of climate change. *Quat. Sci. Rev.* **2006**, *25*, 1383–1400. [CrossRef]

59. Knudsen, K.L.; Seidenkrantz, M.-S.; Kristensen, P. Last interglacial and early glacial circulation in the Northern North Atlantic Ocean. *Quat. Res.* **2002**, *58*, 22–26. [CrossRef]

60. Kaspar, F.; Spangehl, T.; Cubasch, U. Northern hemisphere winter storm tracks of the Eemian interglacial and the last glacial inception. *Clim. Past* **2007**, *3*, 181–192. [CrossRef]

61. Hearty, P.J. Boulder deposits from large waves during the last interglaciation on North Eleuthera Island, Bahamas. *Quat. Res.* **1997**, *48*, 326–338. [CrossRef]

62. Engel, M.; Kindler, P.; Godefroid, F. Speculations on superstorms—Interactive comment on "Ice melt, sea level rise and superstorms: Evidence from paleoclimate data, climate modeling, and modern observations that 2 °C global warming is highly dangerous" by J. Hansen et al. *Atmos. Chem. Phys.* **2015**, *15*, C6270–C6281.

63. Rovere, A.; Casella, E.; Harris, D.L.; Lorscheid, T.; Nandasena, N.A.K.; Dyer, B.; Sandstrom, M.R.; Stocchi, P.; D'Andrea, W.J.; Raymo, M.E. Giant boulders and Last Interglacial storm intensity in the North Atlantic. *Proc. Natl. Acad. Sci. USA* **2017**, *114*, 12144–12149. [CrossRef]

64. Ruban, D.A. Coastal boulder deposits of the Neogene world: A synopsis. *J. Mar. Sci. Eng.* **2019**, *7*, 446. [CrossRef]

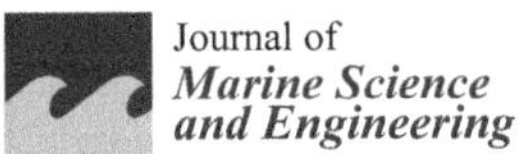
Journal of
*Marine Science and Engineering*

*Article*

# Late Pleistocene Boulder Slumps Eroded from a Basalt Shoreline at El Confital Beach on Gran Canaria (Canary Islands, Spain)

Inés Galindo [1], Markes E. Johnson [2,*], Esther Martín-González [3], Carmen Romero [4], Juana Vegas [1], Carlos S. Melo [5,6,7,8], Sérgio P. Ávila [6,8,9] and Nieves Sánchez [1]

[1] Instituto Geológico y Minero de España, Unidad Territorial de Canarias, c/Alonso Alvarado, 43, 2A, 35003 Las Palmas de Gran Canaria, Spain; i.galindo@igme.es (I.G.); j.vegas@igme.es (J.V.); n.sanchez@igme.es (N.S.)
[2] Department of Geosciences, Williams College, Williamstown, MA 01267, USA
[3] Museo de Ciencias Naturales de Tenerife, Organismo Autónomo de Museos y Centros, C/ Fuente Morales, 1, 38003 Santa Cruz de Tenerife, Spain; MMARTIN@museosdetenerife.org
[4] Departamento de Geografía, Campus de Guajara, Universidad de La Laguna, La Laguna, 38071 Tenerife, Spain; mcromero@ull.esdu.es
[5] Departamento de Geologia, Faculdade de Ciências, Universidade de Lisboa, 1749-016 Lisbon, Portugal; csmelo@fc.ul.pt
[6] CIBIO–Centro de Investigação em Biodiversidade e Recursos Genéticos, InBIO Laboratório Associado, Pólo dos Açores, Universidade dos Açores, Rua da Mãe de Deus, 9500-321 Ponta Delgada, Portugal; sergio.pa.marques@uac.pt
[7] IDL–Instituto Dom Luiz, Faculdade de Ciências, Universidade de Lisboa, 1749-016 Lisbon, Portugal
[8] MPB–Marine Palaeontology and Biogeography lab, Universidade dos Açores, Rua da Mãe de Deus, 9500-321 Ponta Delgada, Portugal
[9] Faculdade de Ciências da Universidade do Porto, Rua do Campo Alegre 1021/1055, 4169-007 Porto, Portugal
[*] Correspondence: markes.e.johnson@williams.edu; Tel.: +1-413-2329

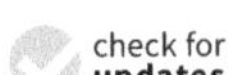
check for **updates**

**Citation:** Galindo, I.s; Johnson, M.E; Martín-González, E.; Romero, C.; Vegas, J.; Melo, C.S; Ávila, S.P; Sánchez, N. Late Pleistocene Boulder Slumps Eroded from a Basalt Shoreline at El Confital Beach on Gran Canaria (Canary Islands, Spain). *J. Mar. Sci. Eng.* **2021**, *9*, 138. https://doi.org/10.3390/jmse9020138

Received: 23 December 2020
Accepted: 19 January 2021
Published: 29 January 2021

**Publisher's Note:** MDPI stays neutral with regard to jurisdictional claims in published maps and institutional affiliations.



**Abstract:** This study examines the role of North Atlantic storms degrading a Late Pleistocene rocky shoreline formed by basaltic rocks overlying hyaloclastite rocks on a small volcanic peninsula connected to Gran Canaria in the central region of the Canary Archipelago. A conglomerate dominated by large, ellipsoidal to angular boulders eroded from an adjacent basalt flow was canvassed at six stations distributed along 800 m of the modern shore at El Confital, on the outskirts of Las Palmas de Gran Canaria. A total of 166 individual basalt cobbles and boulders were systematically measured in three dimensions, providing the database for analyses of variations in clast shape and size. The goal of this study was to apply mathematical equations elaborated after Nott (2003) and subsequent refinements in order to estimate individual wave heights necessary to lift basalt blocks from the layered and joint-bound sea cliffs at El Confital. On average, wave heights in the order of 4.2 to 4.5 m are calculated as having impacted the Late Pleistocene rocky coastline at El Confital, although the largest boulders in excess of 2 m in diameter would have required larger waves for extraction. A review of the fossil marine biota associated with the boulder beds confirms a littoral to very shallow water setting correlated in time with Marine Isotope Stage 5e (Eemian Stage) approximately 125,000 years ago. The historical record of major storms in the regions of the Canary and Azorean islands indicates that events of hurricane strength were likely to have struck El Confital in earlier times. Due to its high scientific value, the outcrop area featured in this study is included in the Spanish Inventory of Geosites and must be properly protected and managed to ensure conservation against the impact of climate change foreseen in coming years.

**Keywords:** coastal storm deposits; storm surge; hydrodynamic equations; upper pleistocene; marine isotope substage 5e; North Atlantic Ocean



*J. Mar. Sci. Eng.* **2021**, *9*, 138. https://doi.org/10.3390/jmse9020138　　　　https://www.mdpi.com/journal/jmse

## 1. Introduction

Evidence for the influence of hurricanes in the northeast Atlantic Ocean during the Late Pleistocene is based on analyses of storm beds preserved on Santa Maria Island in the Azores archipelago [1] and Sal Island in the Cabo Verde archipelago [2]. This line of research follows a growing interest in coastal geomorphology as related to the accumulation of mega-boulders attributed to superstorms or possible tsunami events [3]. Documentation of eroded mega-boulders from rocky shores in Mexico's Gulf of California during the subsequent Holocene [4–6] adds to our knowledge of the physical scope of deposits that can be directly related to historical storm patterns. The Canary Islands in Spain represent seven main islands and numerous islets and seamounts in the North Atlantic located over lithospheric fractures off the northwest coast of Africa, with radiometric dates that vary from 142 Ma to 0.2 Ma along a general east to west axis [7]. At the center of the archipelago, Gran Canaria Island has a volcanic history dominated by subaerial development dating back to 13.7 Ma with successive stages of construction and erosion of the island edifice [8].

The attraction of Gran Canaria for this study is based on a distinctive rocky paleoshore formed by a 3-m thick lava flow exposed laterally along El Confital beach, on the southwest side of La Isleta peninsula, located in the northeastern quarter of Gran Canaria Island. Subaerial flows around several craters preserved atop the Isleta volcanoes exhibit variable Pliocene to Quaternary ages [8], but the basalt shore at El Confital beach is dated to approximately 1 Ma [9,10]. Prior to this contribution, a paleontological study at El Confital beach described an older assemblage of Miocene fossils at one end and a more extensive assemblage of intertidal to shallow subtidal invertebrates confined to the last interglacial epoch attributed to the Eemian Stage, also correlated with Marine Isotope Substage 5e, approximately 125,000 years in age [11]. The goals of this subsequent study include a review of the marine fossil fauna with particular emphasis on species zonation, and with an added emphasis on the extent of encrustations by coralline red algae. Most importantly, this work features extensive analyses of boulders eroded from the paleoshore, among which the Upper Pleistocene biota is preserved. In particular, the physical analyses conducted for clast shape and size are integral to estimates of wave heights based on competing mathematical equations applied to a rocky shoreline composed of joint-bound basalt and hyaloclastite layers.

The range in estimated wave heights extrapolated from average boulder size, contrasted against maximum boulder size at multiple sample sites, suggests that a pattern of repetitious storms is implicated with the coastal erosion of La Isleta and more generally of Gran Canaria Island over an extended period of Late Pleistocene time. The viability of this inference is tested against the historical record of cyclonic disturbances in the eastern North Atlantic Ocean around the Azorean and Canary Islands. Increasing interest in the historical record of storms in this region is available mainly in Spanish-language reports, but also well summarized in the international literature [12]. This approach with reference to historic storms is predicated on a similar analysis of coastal storm beds from the Pleistocene of Santa Maria Island in the Azores [1] as well as coastal boulder beds from the Holocene of Mexico's Baja California [4–6].

## 2. Geographical and Geological Setting

Gran Canaria is the third largest and most centrally located island in the Canary Archipelago, situated 150 km off the northwest coast of Africa (Figure 1a,b). With an area of 1560 km$^2$, the island exhibits a circular map outline with an outer perimeter of about 50 km showing a pattern of ravines radiating from the island center at an elevation of 1949 m above present sea level. La Isleta peninsula lies in the northeastern part of Gran Canaria Island, now linked to it by a sandy isthmus about 200 m in width and around 2 km long (Figure 1b,c). Gran Canaria island emerged during the Miocene and reflects a complex geological evolution including the formation and erosion of several stratovolcanoes. Pliocene-Quaternary mafic fissure eruptions were concentrated along the northeastern half of the island. Volcanic activity on La Isleta began during the early

Pleistocene with the formation of submarine volcanoes more than 1 Ma ago [9,10]. At least two submarine Surtseyan eruptions separated by marine sedimentary rocks have been identified. The later subaerial phase includes several mafic fissure eruptions (Figure 1c). In El Confital Bay, located off the east-southeast part of La Isleta, erosion has dominated since the middle Pleistocene (>152 ka) and marine, aeolian, and colluvial deposits have been deposited overlying the volcanic sequence (Figure 1d). This area was intensively modified by quarry operation, military activities, and expansion of a shantytown. The clearing of the coastal zone after illegal settlement between 1960 and 1995 and its subsequent elimination by civil authorities, resulted in the area of the deposit encompassed by a larger part of the Confital platform being compromised.

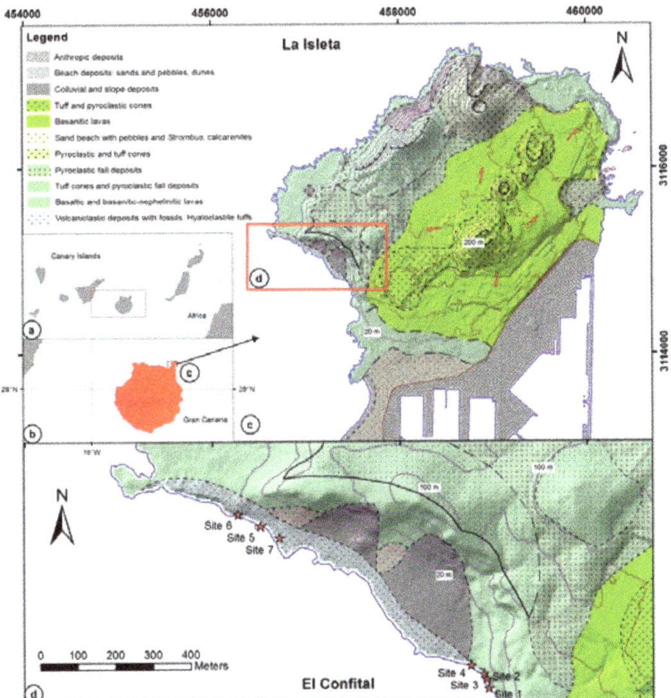



**Figure 1.** Canary Archipelago and features of the centrally located Gran Canaria Island: (**a**) Canary Archipelago with inset showing the position of Gran Canaria; (**b**) central island of Gran Canaria with an arrow pointing to La Isleta on the northern periphery of the island; (**c**) geological map of the volcanic edifice that forms La Isleta with inset box marking El Confital beach (coordinates in UTM m, H28 R); (**d**) map of the kilometer-long beach at El Confital showing seven localities from which sedimentological and paleontological data were drawn for this study. Modified from the 1990 Geological Map by Balcells and Barrera at a scale of 1:25,000 [10].

El Confital bay has a U-shape open to the southwest and is bound on the northeast by an escarpment more than 100 m high. The beach has been developed over a littoral platform (Figure 1c,d) covering approximately 0.1 km$^2$ that consists of highly fractured hyaloclastite tuffs from the submarine stage and volcanoclastic deposits including marine fossils. Parallel to the beach, a basalt cliff line extends parallel to the coast for a broken distance of about a kilometer. Cliffs crop out at an elevation of 1 m in the central part of the area and rise to more than 10 m at opposite ends in the southeast and northwest. In these parts, tuffs are overlain by subaerial lava flows and the cliff reaches a height of more than 10 m.

## 3. Materials and Methods

### 3.1. Data Collection

Gran Canaria was visited in April 2018, when the organizing author was invited to appraise the beach and rocky shoreline at El Confital beach on the east side of Las Palmas de Gran Canaria. The original data for this study was collected in October 2020 from deposits dominated by basalt boulders consolidated within a limestone matrix. Individual basalt clasts from six stations were measured manually to the nearest half centimeter in three dimensions perpendicular to one another (long, intermediate, and short axes). Differentiated from cobbles, the base definition for a boulder adapted in this exercise is that of Wentworth [13] for an erosional clast equal to or greater than 256 mm in diameter. No upper limit for this category is defined in the geological literature [14].

Triangular plots are employed to show variations in clast shape, following the design of Sneed and Folk [15] for river pebbles. In the field, all measured clasts were characterized as sub-rounded and a smoothing factor of 20% was applied uniformity to adjust for the estimated volume first calculated by the simple multiplication of the lengths of the three axes. Comparative data on maximum cobble and boulder dimensions were fitted to bar graphs to show size variations in the long and short axes from one sample to the next. Comparative data on maximum cobble and boulder dimensions were fitted to bar graphs to show size variations in the long and intermediate axes from one sample to the next. The rock density of basalt from the Pleistocene sea cliffs on El Confital beach is based on laboratory analyses in an unpublished PhD thesis that yields a value of 2.84 g/cm$^3$ [16].

### 3.2. Hydraulic Model

Dependent on the calculation of rock density for basalt, a hydraulic model may be applied to predict the force needed to remove cobbles and boulders from a rocky shoreline with joint-bound blocks as a function of wave impact. Basalt is the typical extrusive volcanic rock characteristic of many oceanic islands. Herein, two formulas are applied to estimate the size of storm waves against joint-bounded blocks derived, respectively from Equation (36) in the work of Nott [17] and from an alternative formula using the velocity equations of Nandasena et al. [18], as applied by Pepe et al. [19].

$$Hs = \frac{\left(\frac{\rho_s - \rho_w}{\rho_w}\right) a}{C_1} \tag{1}$$

$$u^2 = \frac{2\left(\frac{\rho_s - \rho_w}{\rho_w}\right) g\, c\, (\cos\theta + \mu_s \sin\theta)}{C_1} \tag{2}$$

where $Hs$ = height of the storm wave at breaking point; $\rho_s$ = density of the boulder (tons/m$^3$ or g/cm$^3$); $\rho_w$ = density of water at 1.02 g/cm$^3$; $a$ = length of the boulder on long axis in cm; $\theta$ is the angle of the bed slope at the pre-transport location (1° for joint-bounded blocks); $\mu_s$ is the coefficient of static friction (=0.7); $C_1$ is the lift coefficient (=0.178). Equation (1) is more sensitive to the length of a boulder on the long axis, whereas Equation (2) is more sensitive to the length of a boulder on the short axis. Therefore, some differences are expected in the estimates of $H_S$.

## 4. Results

### 4.1. Characteristics of the Conglomerate

Two principal facies of Pleistocene conglomerates occur at El Confital. The first is linked to the background cliffs at opposite ends of the bay and the second is related to the open paleo-platform 5 to 10 m distal from the cliffs in the more central part of the beach. Along the higher cliffs, a marine conglomerate fills paleo-channels and forms ridges trending NNE–SSW perpendicular to the shore (Figure 2a,b). The channel conglomerate is well cemented and consists of coarse-grained (pebble to boulder size) clasts embedded in a white matrix of bioclastic calcarenite. The matrix incorporates marine mollusks and calcareous algae. Commonly reaching a cubic meter in size, boulders are generally ellipsoidal to angular in shape and well rounded. The largest boulders exhibit a long axis in excess of 2 m. Locally, the matrix is composed of reddish sands derived from eroded hyaloclastite tuffs (Figure 2c). Neptunian dykes related to the emplacement of this deposit cut into the tuffs and volcanoclastic deposits filled with carbonates and zeolites as well as bioclasts and gravels (Figure 2d). The more distal conglomerate is deposited nearly horizontal across the shore platform. Here, boulders are smaller and more rounded, and the matrix consists of coarse-grained bioclastic sand.

**Figure 2.** Conglomerate slumps above El Confital beach south of La Isleta; (**a**) incision approximately 3.5 m deep in cliff face filled with basal boulders loosened from the adjacent basalt formation (backpack for scale); (**b**) boulder-filled cut in the same cliff face with individual boulders exposed in bas relief (the largest boulder at the bottom has a long axis of approximately 75 cm); (**c**) reddish sands derived from eroded hyaloclastite; (**d**) details showing filling of a neptunian dyke and syndepositional fractures in basalt (tape-measure case for scale).

*4.2. Comparative Variation in Clast Shapes*

Raw data on clast size in three dimensions collected from each of six sample sites are recorded in Appendix A (Tables A1–A6). Regarding shape, points representing individual cobbles and boulders are fitted to a set of Sneed-Folk triangular diagrams (Figure 3a–f). The spread of points across these plots reflects a consistent pattern in the variation of shapes from one sample to another. As few as one to three points fall within the upper triangle in each diagram, which represents an origin from a perfectly cube-shaped endpoint as a joint-bound block of basalt. No more than five points from each sample fall within the lower, right-hand rhomboid in these diagrams. This extreme corner signifies elongated blocks with one super-attenuated axis in relation to two axes that are significantly shorter by 75% or more. The result is a bar-shaped piece that originated as a joint-bound block of basalt. The plot with the greatest number of points in these two extremes is from locality 5 (Figure

3e), from a position closer to the western end of the paleoshore. Numbering between 10 and 15 per sample, the majority of points from each of the six samples fall within the central two rhomboids directly below the top triangle. Such points clustered at the core of a Sneed-Fold diagram are typical of clasts for which two of the three dimensions are closer in value than the length of the third axis. It is notable that no points appear anywhere along the margin of rhomboids on the left side of these diagrams. It is clear that no tendency in shape towards plate-shaped clasts is evident in the data. The general slope of points in agreement from the six plots follows a uniformly diagonal trend from the top to the lower right-hand corner. The trend in distributed points from these plots signifies the rounding of clasts in which two of the dimensions (maximum and intermediate lengths) are more closely matched with the third as an outlier.

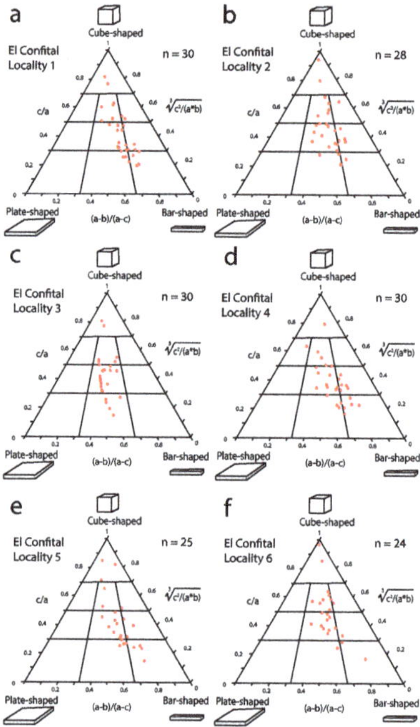

**Figure 3.** Set of triangular Sneed-Folk diagrams used to appraise variations in cobble and boulder shapes sampled along the Upper Pleistocene paleoshore east to west at El Confital beach south of La Isleta: (**a–f**) sample localities 1 to 6 represented, respectively.

### 4.3. Comparative Variation in Clast Sizes

Drawn from original data (Tables A1–A6), clast size is plotted to the best effect on bar graphs as a function of frequency against maximum and intermediate lengths of the two longest axes perpendicular to one another. The dozen graphs plotted (Figure 4a–l) exhibit trends in clast size sorted by intervals of 15-cm, in which the boundary between cobbles and boulders is embedded within the range for clasts between 16 and 30 cm in diameter. The left-hand column (Figure 4a,c,e,g,k) depicts lateral variations in maximum boulder length from the six samples on an east to west transect along the Upper Pleistocene paleoshore. Overall, each of the samples in this dimension is numerically dominated by boulders in contrast to cobbles at ratios from 3:2 and 4:1. Samples from the east end (Figure 4a,c) reflect differences that conform to a normal bell-shaped curve, whereas samples from the west end (Figure 4i,k) are strongly skewed to include a few boulders of extreme size in excess of one meter. In contrast, the right-hand column (Figure 5b,d,f,h,j,l) shows lateral variations in values for intermediate length of clasts. In all but two examples (Figure 4f,j) the number of measurements falling within the interval of 16 to 30 cm exceeds those registered for the same interval as measured for the long axis. This result reflects a general shift in size to smaller frequencies compared to the left-hand column and confirms the diagonal trend in shapes illustrated by the Sneed-Folk diagrams (Figure 4a–f). Skewness that mirrors the inclusion of extra-large clasts is especially evident in the bar graphs at the extreme ends of the paleoshore (Figure 4b,l). The largest basalt clast identified in the entire project registered a long axis of 214 cm, an intermediate axis of 94 cm, and short axis of 50 cm (Table A6).

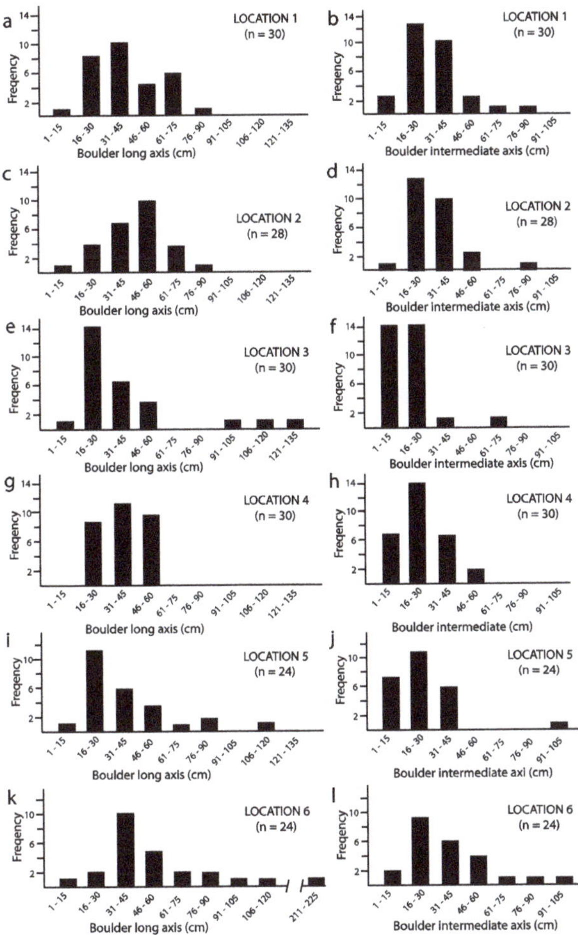

**Figure 4.** Set of bar graphs used to contrast variations in maximum and intermediate boulder axes from six samples at El Confital beach south of La Isleta: (**a**,**b**) bar graphs from locality 1; (**c**,**d**) bar graphs from locality 2; (**e**,**f**) bar graphs from locality 3; (**g**,**h**) bar graphs from locality 4; (**i**,**j**) bar graphs from locality 5; (**k**,**l**) bar graphs from locality 6.

**Figure 5.** Examples of abundant marine fossils from the paleoshore exposed at El Confital beach: (**a**) filling among larger boulders that includes the elbow shell (*Patella aspera*) and barnacles (*Balanus trigonus*) with bottlecap for scale; (**b**) bivalve (*Chama gryphoides*) with pen-tip for scale; (**c**) gastropod (*Gemophos viverratus*) with 3-cm scale; (**d**) gastropod (*Cerithium vulgatum*) with scale bar = 1 cm; (**e**) large colony of gastropods (*Dendropoma cristatum*) with rock hammer for scale.

*4.4. Paleontological Inferences on Water Depth*

A moderately rich molluscan fauna of 42 marine gastropods and eight bivalves from El Confital (Table 1) that also includes barnacles and rhodoliths formed by coralline red algae is maintained in the permanent collections of the Tenerife Science Museum. Overall, these fossils reflect organisms that lived under inter-tidal to shallow subtidal conditions as confirmed by outcrop relationships, where currently the fossil remains are visible cemented in place along small sections of the bay and in ravines that form after heavy rains.

The Pleistocene fauna at El Confital corresponds to a high-energy setting against a rocky shore highlighted by the abundance of mäerl and rhodoliths associated with relatively abundant patelid gastropods (Figure 5a), chamid bivalves (Figure 5b), and other mollusks (Figure 5c,d). Extensive colonies of vermetid gastropods (*Dendropoma cristatum*) are represented as discrete biological clasts incorporated within the conglomerate (Figure 5e). Although extensive shell fragmentation is evident, it is worth noting the high rate of complete shells, which include delicate ornamentation. In the case of pateliform shells, preservation also features evidence for stacking. This taphonomic trait is characteristic of a high-energy regime. At the base of the deposit on the platform at El Confital (Figure 2, locality 7), there is evidence of bioturbation possibly related to the activity of crabs. Also, trace fossils likely related to the activity of polychaets are preserved within the neptunian dykes that are common on the platform.

**Table 1.** Summary species list of epifaunal, infra-intertidal invertebrates from the Upper Pleistocene strata exposed at El Confital beach correlated with Marine Isotope Substage 5e. Extinction is denoted by asterisk.

| Phylum | Class | Species | Phylum | Class | Species |
|---|---|---|---|---|---|
| Mollusca | Gastropoda | *Acanthina dontelei* * <br> *Alvania macandrewi* <br> *A. scabra* <br> *Barleiia unifasciata* <br> *Bolma rugosa* <br> *Bittium reticulatum* <br> *Bursa scrobilator* <br> *Cerithium vulgatum* <br> *Cheilea equestris* <br> *Clanculus berthelotii* <br> *Columbella adansoni* <br> *Conus guanche* <br> *C. pulcher* <br> *Coralliophila meyendorffi* <br> *Dendropoma cristatum* <br> *Diodora gibberula* <br> *Erosaria spurca* <br> *Gemophos viverratus* <br> *Gibbula candei* <br> *Gibbula* sp. <br> *Haliotis tubeculata* <br> *Littorina littorea* <br> *Luria lurida* <br> *Manzonia crassa* <br> *Marginella glabella* <br> *Mitra cornea* <br> *Monoplex parthenopeus* <br> *Naria spurca* <br> *Patella aspera* <br> *P. candei* <br> *P. crenata* <br> *P. piperata* <br> *Phorcus atratus* <br> *P. sauciatus* <br> *Pusia zebrina* <br> *Stramanita haemastoma* <br> *Tectarius striatus* <br> *Thylacodes arenarius* <br> *Vermetus triquetrus* <br> *Vermetus* sp. <br> *Vexillum zebrinum* <br> *Zebina vitrea* | Mollusca | Bivalvia | *Barbatia barbata* <br> *Bractechlamys corallinoides* <br> *Cardita calyculata* <br> *Chama gryphoides* <br> *Ctema decussata* <br> *Glycymeris glycymeris* <br><br> *Pecten* sp. <br> *Venus verrucosa* |
| | | | Arthropoda | Cirripedia | *Balanus trigonus* |

In addition to preservation of whole but also fragmented rhodolith debris (Figure 6a,d) that signify the remains of a Pleistocene mäerl bed at El Confital, deposits are notable for the conglomerate consisting of mixed basalt cobbles and large boulders that exhibit erosional smoothing. The growth of coralline red algae in thin layers encrusted around and among these basalt clasts is widespread (Figure 6). In some examples, algal crusts are localized and fail to completely surround individual cobbles leaving some parts free as viewed profile (Figure 6a). This scenario implies that some clasts were only partially coated by crustose algae elsewhere and subsequently were transferred to the conglomerate. In other examples, thick growth of algal crusts completely fills voids between boulders and small cobbles fixed in between (Figure 6b). This suggests an alternative scenario in which algal growth occurred perhaps on the outer margin of the conglomerate after its mass accumulation. Crustose red algae also are found in patches attached to basalt boulders exposed in three-dimensional relief (Figure 6c).

**Figure 6.** Examples of basalt cobbles and boulders heavily encrusted by coralline red algae: (**a**) space between two adjacent boulders filled by algal-encrusted cobbles and pebbles (scale bar = 2.5 cm); (**b**) close-up view of thick algal encrustations in the space among basalt clasts (bottle cap for scale); (**c**) hat-size basalt boulder with remnants of coralline red algae (clasp on hat string = 3 cm).

### 4.5. Storm Intensity as a Function of Estimated Wave Height

Clast sizes and maximum boulder volumes drawn from the six field localities are summarized in Table 1, allowing for direct comparison of average values for all clasts, as well as values for the largest clasts in each sample based on Equations (1) and (2) derived from the work of Nott [16] and Pepe et al. [19].

The Nott formula [17] shown in Equation (1) yields an average wave height of 4.5 m for the extraction of joint-bound blocks from basalt sea cliffs exposed at El Confital beach and their subsequent transfer as slump boulders in samples 1 to 6. A much larger value for a wave height of 11 m is derived from the average of the largest single blocks of basalt recorded from the six locations. More sensitive to clast length from the short axis, the more sophisticated Equation (2) from Pepe et al. [19] yields values that are consistently lower for the estimated average wave height and estimated maximum wave height. The difference between the two calculations for estimated average wave height as well as estimated maximum wave height is, however, very small at 30 cm. Notably, the value for average maximum wave height derived from Equation (1) from Nott [17] is 2.5 times as high as the value found for overall average wave height using the same formula. Essentially the same factor applies to the slightly lesser values derived from Equation (2) according to Pepe et al. [19]. Clearly, the hydraulic pressure with extreme wave impact is necessary to loosen and budge the largest fault-bound blocks of basalt in the Pleistocene cliff line, as represented by the enormous block at locality 6 estimated to weigh 2.25 metric tons (Table A6). The essential issues under consideration in the following sections pertain to the singularity of a lone event of extreme magnitude as opposed to the repetition of many, but less energetic events in the shaping of the Pleistocene boulder slumps at El Confital beach.

## 5. Discussion

### 5.1. Integration of Paleontological and Physical Data

Boulders derived from horizontal layers of joint-bound basalt that originated as a subaerial flow about 1 million years ago at El Confital are estimated to have undergone wear that resulted in a 20% reduction in volume from more cubic or bar-shaped blocks (Figure 4) due to mutual friction under wave shock and subsequent erosion that smoothed sharp corners. Pleistocene fossils incorporated within the resulting conglomerate (Table 1, Figure 5) reflect an age correlated with Marine Isotope Stage 5e during the last interglacial epoch [11,20–23].

The assemblage represents a high-diversity, intertidal to shallow subtidal fauna dominated by mollusks that thrived on a basalt shelf on the southern margin of a small volcanic edifice. Except for those few limited to the MIS5e, most of the species listed in Table 1 continue to inhabit the contemporary coasts of the Canary Islands. These taxa are characterized by a preference for rocky or mixed littoral bottoms (sandy with rocky clasts of different sizes) up to a depth of about 3 m within the intertidal zone. The various species occupy different niches, such as crevices or intertidal pools or beneath rocks, which provide protection from the intense northwesterly waves that dominate the shores of the Canary Islands. In the case of El Confital today, there also occurs a wide rocky intertidal flat with little slope. In these shallows, the organisms are grouped in parallel bands or remain associated with pools at low tide, depending on their ability to adapt to environmental factors such as desiccation, temperature, salinity, and water agitation that condition life in that environment [23]. The intertidal shallows are home to a high number of marine organisms, highlighted by the dominant populations of pateliform and trochid gastropods.

Some of the larger clasts and boulders are encrusted with crustose red algae exhibiting rinds in excess of a centimeter in thickness (Figure 6). In places, surviving patches of red algae retain a rose coloration typical of coralline red algae (Figure 6c), but it may be due to inorganic discoloration. In addition to platy red algae cemented directly onto cobbles and boulders, whole rhodoliths and the debris of broken rhodoliths occur in pockets scattered throughout the conglomerate (Figure 5a). Some faunal elements may have lived within the interstices of adjacent boulders after the conglomerate was formed during slump events. Platy red algae are concentrated unevenly on the sides of cobbles and smaller boulders (Figure 6a), leaving other faces vacant. Some open spaces between adjacent boulders appear to have been filled by the continual growth of crustose red algae (Figure 6b). On the other hand, rhodoliths potentially composed of the same species of coralline red algae in unattached growth forms expressed by spherical shapes would have expired due to a lack of mobility.

A proper survey has yet to be undertaken to identify the genera and possible range of species belonging to coralline red algae that encrust cobbles and boulders in the Upper

Pleistocene deposits at El Confital. Such a study necessarily entails the collection of samples for the making of thin sections transversely through crusts in order to identify features diagnostic at least on a genus level. Until such time, the best that can be said is that the present-day distribution of marine algae is widespread throughout the Canary Islands and keyed to habitats around individual islands in the Canary archipelago. Among the Corallinaceae known to be specific to Gran Canaria island, at least five genera are present, including *Hydrolithon*, *Lithophyullum*, *Lithoporella*, *Mesophylum*, and *Neogoniolithon*. Among these, *Lithophyllum* is the most diverse with four species locally attributed to that genus around the island [24]. From a paleoecological point of view, what is most telling about the boulder beds at El Confital at this stage of investigation is that they came to reside in shallow water within the upper photic zone consistent with the associated fossil fauna.

Overall, the combination of paleontological and physical evidence points to an open shelf setting on the margin of a small volcano around which a distinctly shallow-water biota thrived prior to interruption by storm events that eroded a series of parallel channels perpendicular to the strike of the paleoshore. These submarine gullies define the depositional space in which the Upper Pleistocene conglomerate was accommodated and preserved (Figure 3a,b).

Based on the application of two competing mathematical models that consider different dynamics [17,19], the estimated average height of storm waves that broke onto the south shore of La Isleta are remarkably similar in the range between 4.2 and 4.5 m (Table 2). However, the largest half-dozen boulders sampled from six study sites yield a much higher average estimated height of storm waves between 10.8 and 11.1 m (Table 2). Under any circumstances, such results would represent extremely large waves. These numbers represent clear outliers in the data, although based on boulders of extraordinary size and weight up to 2.25 metric tons. It may be that lesser waves were instrumental in gradually loosening the biggest basalt blocks from a joint-bound condition to a point where gravity slid them into nearby channels enlarged by multiple storm events. Rare hurricane events are more likely to have generated storm waves on the order of 6 to 8 m that impacted the Pleistocene rocky shore at El Confital. An entirely different set of equations is used to estimate the landward onrush of water due to tsunami events, but the complete absence of basalt boulders on the slopes of La Isleta (Figure 2) mitigates against this scenario.

**Table 2.** Summary data from Appendix A (Tables A1–A6) showing maximum bolder size and estimated weight compared to the average values for sampled boulders from each of the transects together with calculated values for wave heights estimated as necessary for boulder-beach mobility. Abbreviations: EAWH = estimated average wave height, EMWH = estimated maximum wave height.

| Confital Locality | Number of Samples | Average Boulder Volume (cm³) | Average Boulder Weight (kg) | EAWH (m) Nott [17] | EAWH (m) Pepe et al. [19] | Max. Boulder Volume (cm³) | Max. Boulder Weight (kg) | EMWH (m) Nott [17] | EMWH (m) Pepe et al. [19] |
|---|---|---|---|---|---|---|---|---|---|
| 1 | 30 | 28,065 | 79.7 | 4.5 | 3.9 | 212,173 | 602.6 | 7.4 | 12.7 |
| 2 | 28 | 9786 | 116.1 | 4.8 | 5.3 | 298,742 | 848.4 | 9.9 | 10.4 |
| 3 | 30 | 22,870 | 64.9 | 4.1 | 3.5 | 252,000 | 713.7 | 10 | 10.2 |
| 4 | 30 | 14,823 | 43.1 | 3.9 | 3.3 | 50,540 | 143.4 | 5.9 | 6.8 |
| 5 | 24 | 34,267 | 77.8 | 4.2 | 3.5 | 238,853 | 678.3 | 11.9 | 8.6 |
| 6 | 24 | 84,792 | 239.7 | 5.5 | 5.6 | 804,640 | 2258 | 21.5 | 16.1 |
| Average | 27.66 | 32,434 | 103.5 | 4.5 | 4.2 | 309,491 | 874.1 | 11.1 | 10.8 |

*5.2. Inference from Historical Storms in the North Atlantic*

Given their geographic location, archipelagos located north of Cabo Verde off the northwest coast of Africa are likely to be impacted by high-energy storms [25]. The Azores Archipelago is struck by high-energy storms with a frequency every seven years [26], causing several shipwrecks in the harbor of Ponta Delgada on São Miguel (Figure 7a). More recently, the passage of Hurricane Lorenzo in October 2019 caused the destruction of several piers among the islands, as well as the near disappearance of Lajes das Flores harbor (Figure 7b). The Canary Islands are no exception. The high-energy events that affect the islands have caused considerable damage [12] and even fatalities (Figure 7c,d).

**Figure 7.** Evidence of destruction caused by storms in the Azorean and Canary islands; (**a**) Shipwrecks in Ponta Delgada harbor (São Miguel, Azores) after the December 1996 storm (photo by João Brum); (**b**) Destruction of the pier of Lajes das Flores (Flores Island, Azores) after the passage of Hurricane Lorenzo in October 2019 (Flores Island, Azores) after the passage of Hurricane Lorenzo in October 2019 (*Journal Açores* 9, 2019); (**c**) Shipwreck beached in Gran Canaria after the November 1968 storm (*Efemérides Meteorlógicas de Canárias*, 2018); (**d**) destruction caused by the April 1970 storm that hit Gran Canari (*Efemérides Meteorlógicas de Canárias*, 2018).

Historical records (Table 3) are scarce in terms of available wave-height information. However, in some cases inferences can be made: the events of 21 February 1966 (10 to 12 m), 23 November 1968 (10 m), and 21 April 1970 (8 m). These inferences suggest that storm waves can reach considerable values in the Canary Islands. Empirical models for wave modulation about the island of Gran Canaria indicate average wave height values between 5.22 and 5.58 m [27,28]. Such wave heights are compatible with the maximum average values estimated for boulders summarized for location 6 (Table 2) with averages based on Equation (1) [17] of 5.5 m and Equation (2) [19] of 5.6 m. The location of El Confital beach within the bay also is pertinent. The development of the volcanic edifice of La Isleta (Figure 1c) played an important role in protecting the area. Although consideration of the effect and action of waves before the Last Interglacial Epoch is possible, the formation of a large sandy isthmus since that time connected the La Isleta to the larger island of Gran Canaria, making the area even more sheltered from storm events. No direct evidence is observable at El Confital, but it is necessary to remember that the Canary Islands are subject to tsunami waves resulting from earthquakes like the wider regional event of 1755 that destroyed Lisbon, as well as volcanic flank collapse on the home islands.

**Table 3.** Coastal disturbances between 1755 and 2009 that affected Gran Canaria and islands elsewhere in the Canary Archipelago.

| Order | Date | Type Event | Locations | Scale of Destruction |
|---|---|---|---|---|
| 1 | 25 January 1713 | major storm | Gran Canaria and Tenerife | Major destruction, overflow in Vega de Arucas |
| 2 | 1 November 1755 | tsunami | Tenerife, Gran Canaria, and Lanzarote | Destruction in coastal areas due to Lisbon earthquake |
| 3 | 6 January 1766 | major storm | Gran Canaria | Torrential rain; lahaar in Agüimes |
| 4 | 7–8 November 1826 | major storm | all islands | Damage throughout the Canaries and loss of life |
| 5 | 7 January 1856 | hurricane | Tenerife and El Hierro | 50 buildings destroyed including the harbor and pier; two fatalities |
| 6 | 13 February 1875 | major storm | Las Palmas de Gran Canaria | Damage to ships in the harbor, one fatality |
| 7 | 16 January 1895 | major storm | Gran Canaria | Overflow at E Confital, house destruction |
| 8 | 25–26 December 1898 | major storm | Gran Canaria | Damage to pier and docked ships |
| 9 | 3 March 1903 | major storm | Gran Canaria | No reported damage |
| 10 | 16 December 1903 | major storm | Gran Canaria | Raging sea; one fatality |
| 11 | 2 February 1904 | major storm | Gran Canaria | Disappearance of sand from El Confital |
| 12 | 28–29 October 1905 | major storm | Gran Canaria | Streets of Las Palmas flooded; one fatality |
| 13 | 17 January 1910 | major storm | Gran Canaria | Pier damage |
| 14 | 30 January 1910 | major storm | Gran Canaria | Overflow in Las Palmas; one fatality |
| 15 | 26–27 December 1910 | major storm | All islands | Las Palmas harbor closed to traffic |
| 16 | 1912 | major storm | Gran Canaria | Overflow in Las Palmas |
| 17 | 7–9 February 1912 | major storm | Gran Canaria | Homes in Las Canteras destroyed |
| 18 | 22 November 1913 | major storm | Gran Canaria | Homes in Las Canteras damaged |
| 19 | 26 November 1915 | major storm | Gran Canaria | Harbor closure |
| 20 | 19 November 1933 | major storm | Gran Canaria | Wave destruction; one fatality at El Confital |
| 21 | 12–13 December 1957 | major storm | Gran Canaria | Major destruction to Las Nieves harbor |
| 22 | 10 September 1961 | major storm | Gran Canaria | one fatality at Las Salinas de El Confital |
| 23 | 21 February 1966 | major storm | Gran Canaria | Wave heights between 10 and 12 m |
| 24 | 23–26 November 1968 | major storm | Tenerife and Gran Canaria | Wave height of 10 m in Tenerife; wind speed of 118 km/h in Gran Canaria |
| 25 | 21 April 1970 | major storm | Gran Canaria | Wave height of 8 m; wind speed over 90 km/h |
| 26 | 18 February 1971 | major storm | Gran Canaria | Damage in El Confital |
| 27 | February 1989 | major storm | Gran Canaria | Overflow; home damage; closure of Puerto de La Luz harbor |
| 28 | 28–29 November 2005 | major storm DELTA | all islands | Damage to harbors and beaches |
| 29 | 21 December 2009 | major storm | Gran Canaria | Overflow; precipitation levels of 130 mm/h; damage to homes |

### 5.3. Comparison with Coastal Boulder Deposits Elsewhere

Pleistocene and Holocene deposits formed by cobbles and boulders are widely distributed all around the world [3], but studies in coastal geomorphology seldom consider density as related to parent rock types when investigating the range of wave heights necessary for their development as eroded boulders. Application of mathematical formulae such as Equation (1) from Nott [17] to estimate storm wave height has been applied previously to coastal boulder deposits throughout Mexico's Gulf of California, including those formed by limestone, rhyolite, and andesite clasts [4–6]. Extension of this work to include Equation (2) as derived from Pepe et al. [19] also has been applied to coastal basalt deposits in the Azores [1]. The variation in mass among these rock types ranges from 1.86 gm/cm$^3$ for limestone, 2.16 gm/cm$^3$ for the rhyolite, 2.55 gm/cm$^3$ for andesite, and 3.0 gm/cm$^3$ for basalt.

Among the largest Holocene boulders treated in these studies are those derived from rhyolite shores on the Gulf of California yielding mega-boulders with a calculated weight of 4.3 metric tons requiring an estimated wave height of 16.8 m to shift from the parent rocky shore [5]. In this particular case, the estimated wave height from storms in the Gulf of California is commensurate with the wave height formulated for the largest basalt boulder recorded anywhere on El Confital beach at study site 6 (Table 2). The recent history of hurricanes in the Gulf of California filmed under direct observation, confirms wave heights at least half that size against contemporary rhyolite sea cliffs [5].

The island of Santa Maria in the Azores has yielded a study comparing present-day and Upper Pleistocene basalt boulders [1] most applicable to the present study at El Confital beach in Gran Canaria. The largest Pleistocene boulder recorded in that study is smaller than many of the typical basalt boulders from El Confital, but still exemplified an estimated wave height of 8 m necessary for its emplacement. The largest boulder from a modern deposit on Santa Marine Island was calculated to require a wave height of 6.4 m for its emplacement [1].

Technically, hurricanes have a tropical to subtropical source dependent on high ocean-surface water temperatures and excessive air moisture [29]. However, major storms of hurricane intensity also occur in Arctic latitudes, where contemporary and Holocene boulder deposits occur along the coast of Norway. Small boulders formed by low-grade chromite ore with a density of 3.32 g/cm$^3$ are described from Holocene deposits on Norway's Leka Island that imply wave heights as much a 7 m for their emplacement [30].

*5.4. Notes on the Geoheritage of El Confital Beach*

El Confital beach takes its name from the former abundance of "confites" (candies in English) on the beach, a popular name given to rhodoliths throughout the Canary Islands for their white color and ball-like shape. Notably, the study area at El Conital beach is richly fossiliferous as known since the visit by Charles Lyell in 1854 to Gran Canaria island [31]. Historically, deposits with fossil rhodoliths, as well as other carbonates were exploited massively to manufacture lime, due to the lack of this resource in the Canary Islands. This industry has led to the loss of essential paleontological paleoecologial, and taphonomical information. The outcrop at El Confital was chosen as a Geosite (site of geological interest, acronym LIG in Spanish) [32,33] and is included in the Inventory of Geosites of the Canary Islands, carried out by project LIGCANARIAS [11,34], due to its high scientific value that represents an area where different types of geological heritage are combined. Stratigraphic sequence is the central feature around which others are related including paleontology, sedimentology, and geomorphology [34–37]. The volcano-sedimentary sequence at El Confital reaches a maximum level of 200 m above sea level in which characteristics of the geological evolution of Gran Canaria island are represented.

As shown in this study, the significance of El Confital is magnified as an example of an accumulation zone of basaltic boulders of different sizes that denote high-energy events essential to understanding the impact of storms and hurricanes in the island groups of the North Atlantic. Apart from the materials belonging to the last interglacial maximum (MIS5e) described in this paper, El Confital includes a range of other features represented by submarine and subaerial basaltic deposits and hyaloclastites (peperites) together with marine sands and conglomerates, aeolian sand dunes, and colluvial deposits [35].

For all its value as a Geosite with high scientific, educational, and touristic value [32,33,35,36], El Confital beach is extremely fragile and vulnerable to human impact and climate change. Therefore, it remains necessary to adopt a management plan that ensures its regulatory protection in the short- and intermediate-term [37]. Although adjacent to the Bahia del Confital Special Conservation Area, Community Interest Area, as well as La Isleta Marine Area, and the Protected Landscape of La Isleta, it is urgent that the Geosite attain an effective geoconservation plan sanctioned by the regional government of the Canary Islands.

## 6. Conclusions

Study of the cobble-boulder deposits at Playa El Confital offers insights based on mathematical equations for estimation of Late Pleistocene wave heights from super-storms in the same region:

- Consolidated cobble-boulder deposits preserved in multiple samples from Upper Pleistocene strata exhibit evidence of major slumps from a rocky shoreline formed by basalt flows during a prior stage of development related to the small volcanic peninsula of La Isleta.
- Preserved in conglomerate deposits that are well cemented, the average estimated volume and weight of individual basalt boulders from a total of 166 samples suggest wave heights between 4.2 and 4.5 m responsible for their derivation from an adjacent stratiform and joint-bound body of parent rock. The largest basalt boulder from all six sample sites is estimated to weigh 2.25 metric tons and may have been moved by a wave of extraordinary height around 10 m. Alternately, smaller waves may have gradually loosened this block from its parent body until the force of gravity entrained it within the conglomerate.
- Often ellipsoidal to angular in shape but typically well rounded, the degree of wear to which individual boulders were subjected implies the action of multiple storm events that also gradually enlarged the size of the channels in which the conglomerates were entrained. Ellipsoidal shapes were governed by the spacing of vertical joints in the parent basalt flow.
- An associated marine biota consisting of diverse mollusks dates the conglomerates to an age consistent with Marine Isotope Stage 5e, equivalent to the Eemian Stage during the last interglacial epoch. The biota also includes rhodoliths formed by coralline red algae growing in spherical forms unattached to the seabed, as well as abundant evidence of platy red algae encrusted directly onto many boulders. Much the same biota lives the present-day embayment at El Confital and represents an inter-tidal to very shallow subtidal habitat.
- Historical records from major storm events that impacted the Canary and nearby Azorean islands confirm that wave heights in the range of those predicted by mathematical models for the erosion of the El Confital conglomerates are reasonable for erosion of all but perhaps the largest boulders entrained, therein.
- Given the importance of geoheritage at El Confital beach and the boulder deposits described in this paper, it remains necessary to implement an adequate management plan against human impact and climate change.

**Author Contributions:** M.E.J. initiated the project as a contribution to the Special Issue in the Journal of Marine Sciences and Engineering devoted to "Evaluation of Boulder Deposits Linked to Late Neogene Hurricane Events." I.G. and N.S. collected the extensive raw data on bolder dimensions at the study site. All co-authors contributed various parts of the text, with I.G. and N.S. contributing material on geologic background and characteristics of the conglomerate deposit. E.M.-G. was responsible for the paleontological and paleoecological description. M.E.J. and S.P.Á. compiled the statistics on wave heights. C.R. and C.S.M. researched and summarized existing historical records on the region's major storms. J.V. contributed the summary on geoheritage. All authors have read and agreed to the published version of the manuscript.

**Funding:** This research received funding from the Canarian Agency for Research, Innovation, and Society of Information (ACIISI) under the Government of the Canary Islands through the project ProID2017010159. C.S. Melo is the recipient of a PhD scholarship M.3a/F/100/2015 from FRCT/Açores 2020 from the Regional Fund for Science and Technology (FRCT) and benefitted from S.P.Á. "Projecto Exploratório" IF/00465 from the Foundation for Science and Technology (FCT). He also expresses appreciation for help from the FEDER through the Operational Program for Competitiveness Factors–COMPETE; by FCT under projects UID/BIA/50027/2013 and POCI-01-01-0145-FEDER-006821; DRCT1.1 project a/005/Function-C-/2016 (CIBIO-A) of the FRCT.

**Acknowledgments:** To be added following the review process.

**Conflicts of Interest:** The authors declare no conflict of interest.

## Appendix A

**Table A1.** Quantification of clast size, volume, and estimated weight from location 1 at the east end of Playa El Confital. The density of basalt at 2.84 g/cm$^2$ is applied uniformly in order to calculate wave height for each boulder on the basis of competing equations. Abbreviation: EWH = estimated wave height.

| Sample | Long Axis (cm) | Intermediate Axis (cm) | Short Axis (cm) | Volume (cm$^3$) | Adjust to 80% | Weight (kg) | EWH Nott [17] (m) | EWH Pepe et al. [19] (m) |
|---|---|---|---|---|---|---|---|---|
| 1 | 28.5 | 15 | 5.5 | 2351 | 1881 | 5.3 | 2.9 | 1.2 |
| 2 | 41 | 34 | 25 | 34,850 | 27,880 | 79.2 | 4.1 | 5.7 |
| 3 | 30 | 25 | 24 | 18,000 | 14,400 | 40.0 | 3.0 | 5.4 |
| 4 | 16 | 11 | 7 | 1232 | 986 | 2.8 | 1.6 | 1.6 |
| 5 | 29 | 21 | 13.2 | 8039 | 6431 | 18.2 | 2.9 | 3.0 |
| 6 | 44 | 19 | 11 | 9196 | 9357 | 26.6 | 4.4 | 2.5 |
| 7 | 28 | 25 | 9.5 | 6650 | 5320 | 15.0 | 2.8 | 2.2 |
| 8 | 58 | 53.5 | 17 | 52,751 | 42,201 | 119.9 | 5.8 | 3.8 |
| 9 | 54 | 46 | 25.5 | 63,342 | 50,674 | 143.9 | 5.4 | 5.8 |
| 10 | 68.5 | 34 | 20 | 46,580 | 37,264 | 105.8 | 6.9 | 4.5 |
| 11 | 37 | 23 | 22 | 18,722 | 14,978 | 42.5 | 3.7 | 5.0 |
| 12 | 35 | 34 | 18 | 21,420 | 17,136 | 48.7 | 3.5 | 4.1 |
| 13 | 24 | 16 | 8 | 3072 | 2458 | 7.0 | 2.4 | 1.8 |
| 14 | 14 | 10 | 7.5 | 1050 | 840 | 2.4 | 1.4 | 1.7 |
| 15 | 74 | 64 | 56 | 265,216 | 212,173 | 602.6 | 7.4 | 12.7 |
| 16 | 38 | 23 | 18 | 15,732 | 12,586 | 35.7 | 3.8 | 4.1 |
| 17 | 63 | 36 | 22 | 49,896 | 39,917 | 113.4 | 6.3 | 5.0 |
| 18 | 46 | 24 | 23 | 25,392 | 20,314 | 57.7 | 4.6 | 5.2 |
| 19 | 53 | 34 | 16 | 28,832 | 23,066 | 65.5 | 5.3 | 3.6 |
| 20 | 21 | 18 | 13 | 4914 | 3931 | 11.0 | 2.1 | 2.9 |
| 21 | 45 | 21 | 10 | 9450 | 7560 | 21.5 | 4.5 | 2.3 |
| 22 | 41 | 20.5 | 13 | 10,927 | 8741 | 24.8 | 4.1 | 2.9 |
| 23 | 75 | 45 | 15 | 50,625 | 40,500 | 115.0 | 7.5 | 3.4 |
| 24 | 45 | 41 | 24 | 44,280 | 35,424 | 100.6 | 4.5 | 5.4 |
| 25 | 72 | 36 | 23 | 59,616 | 47,693 | 135.4 | 7.2 | 5.2 |
| 26 | 71 | 34 | 18 | 43,452 | 34,762 | 98.7 | 7.1 | 4.1 |
| 27 | 88 | 59 | 23 | 119,416 | 95,533 | 271.3 | 8.8 | 5.2 |
| 28 | 43.5 | 26 | 13 | 14,703 | 11,762 | 33.4 | 4.4 | 2.9 |
| 29 | 23 | 18 | 10 | 4140 | 3312 | 9.4 | 2.3 | 2.3 |
| 30 | 40 | 33.5 | 12 | 16,080 | 12,864 | 36.5 | 4.0 | 2.7 |
| Average | 44.85 | 30 | 17.5 | 34,998 | 28,065 | 79.7 | 4.5 | 3.9 |

**Table A2.** Quantification of clast size, volume, and estimated weight from location 2 at the east end of Playa El Confital. The density of basalt at 2.84 g/cm$^2$ is applied uniformly in order to calculate wave height for each boulder on the basis of competing equations. Abbreviation: EWH = estimated wave height.

| Sample | Long Axis (cm) | Intermediate Axis (cm) | Short Axis (cm) | Volume (cm$^3$) | Adjust to 80% | Weight (kg) | EWH Nott [17] (m) | EWH Pepe et al. [19] (m) |
|---|---|---|---|---|---|---|---|---|
| 1 | 99 | 82 | 46 | 373,428 | 298,742 | 848.4 | 9.9 | 10.4 |
| 2 | 71 | 48 | 27 | 92,016 | 73,613 | 209.1 | 7.1 | 6.1 |
| 3 | 27 | 19 | 10 | 5130 | 4104 | 11.7 | 2.7 | 2.3 |
| 4 | 55 | 27 | 25 | 37,125 | 29,700 | 10.5 | 5.5 | 5.7 |
| 5 | 33 | 16 | 8 | 4224 | 3379 | 9.6 | 3.3 | 1.8 |
| 6 | 46.5 | 45 | 43 | 89,978 | 71,982 | 204.4 | 4.7 | 9.7 |
| 7 | 62 | 24 | 24 | 35,712 | 28,570 | 81.1 | 6.2 | 5.4 |
| 8 | 56 | 30 | 20 | 33,600 | 26,880 | 76.3 | 5.6 | 4.5 |
| 9 | 58 | 51 | 35 | 103,530 | 82,824 | 235.2 | 5.8 | 7.9 |
| 10 | 19 | 15 | 7.5 | 2138 | 1710 | 4.9 | 1.9 | 1.7 |
| 11 | 47.5 | 42.5 | 37.5 | 75,703 | 60,563 | 172 | 4.8 | 8.5 |
| 12 | 46 | 30 | 54 | 74,520 | 59,616 | 169.3 | 4.6 | 12.2 |
| 13 | 64 | 42 | 36 | 96,768 | 77,414 | 219.9 | 6.4 | 8.2 |
| 14 | 47 | 29 | 24 | 32,712 | 26,170 | 74.3 | 4.7 | 5.4 |
| 15 | 27 | 18.5 | 7.5 | 3746 | 2997 | 8.5 | 2.7 | 1.7 |
| 16 | 48 | 35 | 23.5 | 39,480 | 31,584 | 89.7 | 4.8 | 5.3 |
| 17 | 77 | 46.5 | 25 | 89,513 | 71,610 | 203.4 | 7.7 | 5.7 |
| 18 | 42 | 41 | 20 | 34,440 | 27,552 | 78.2 | 4.2 | 4.5 |
| 19 | 33 | 31.5 | 13.5 | 14,033 | 11,227 | 31.9 | 3.3 | 3.1 |
| 20 | 43 | 40 | 27.5 | 47,300 | 37,840 | 107.5 | 4.3 | 6.2 |
| 21 | 36 | 34 | 14 | 17,136 | 13,709 | 38.9 | 3.6 | 3.2 |
| 22 | 19 | 17 | 13 | 4199 | 3359 | 9.5 | 1.9 | 2.9 |
| 23 | 69 | 24 | 21 | 34,776 | 27,821 | 79.0 | 6.9 | 4.8 |
| 24 | 56 | 31 | 24 | 41,664 | 33,331 | 94.7 | 5.6 | 5.4 |
| 25 | 30 | 17 | 15 | 7650 | 6120 | 17.4 | 3.0 | 3.4 |
| 26 | 31 | 26 | 18 | 14,508 | 11,606 | 33.0 | 3.1 | 4.1 |
| 27 | 53 | 32 | 16.5 | 27,984 | 22,387 | 63.6 | 5.3 | 3.7 |
| 28 | 43.5 | 31.5 | 22 | 30,146 | 24,116 | 68.5 | 4.4 | 5.0 |
| Average | 49 | 33 | 23.5 | 12,213 | 9786 | 116.1 | 4.8 | 5.3 |

**Table A3.** Quantification of clast size, volume, and estimated weight from location 3 at the east end of Playa El Confital. The density of basalt at 2.84 g/cm$^2$ is applied uniformly in order to calculate wave height for each boulder on the basis of competing equations. Abbreviation: EWH = estimated wave height.

| Sample | Long Axis (cm) | Intermediate Axis (cm) | Short Axis (cm) | Volume (cm$^3$) | Adjustto 80% | Weight (kg) | EWH Nott [17] (m) | EWH Pepe et al. [19] (m) |
|---|---|---|---|---|---|---|---|---|
| 1 | 53.5 | 26 | 24 | 33,384 | 26,707 | 75.8 | 5.4 | 5.4 |
| 2 | 45 | 29 | 22 | 28,710 | 22,968 | 65.2 | 4.5 | 5.0 |
| 3 | 31 | 26 | 17 | 13,702 | 10,962 | 31.1 | 3.1 | 3.8 |
| 4 | 50 | 30 | 23 | 34,500 | 27,600 | 78.3 | 5.0 | 5.2 |
| 5 | 33 | 25 | 17 | 14,025 | 11,220 | 31.9 | 3.3 | 3.8 |
| 6 | 100 | 70 | 45 | 315,000 | 252,000 | 713.7 | 10.0 | 10.2 |
| 7 | 22 | 18 | 17 | 6732 | 5386 | 15.3 | 2.2 | 3.8 |
| 8 | 25 | 13 | 11.5 | 3738 | 2990 | 8.5 | 2.5 | 2.6 |
| 9 | 27.5 | 12 | 10 | 3300 | 2640 | 7.5 | 2.8 | 2.3 |
| 10 | 52 | 23 | 14 | 16,744 | 13,395 | 38.0 | 5.2 | 3.2 |
| 11 | 17.5 | 20 | 20 | 7000 | 5600 | 15.9 | 1.8 | 4.5 |
| 12 | 44 | 11 | 11 | 5324 | 4259 | 12.1 | 4.4 | 2.5 |
| 13 | 47 | 19 | 19 | 16,967 | 13,574 | 38.6 | 4.7 | 4.3 |
| 14 | 35 | 11 | 11 | 11,235 | 3388 | 9.6 | 3.5 | 2.5 |
| 15 | 145 | 21 | 21 | 63,945 | 51,156 | 145.3 | 14.5 | 4.8 |
| 16 | 26 | 10 | 10 | 2600 | 2080 | 5.9 | 2.6 | 2.3 |
| 17 | 13.5 | 7 | 7 | 662 | 529 | 1.5 | 1.4 | 1.6 |
| 18 | 32 | 6 | 6 | 1152 | 922 | 2.6 | 3.2 | 1.4 |
| 19 | 52.5 | 18 | 18 | 17,010 | 13,608 | 38.6 | 5.3 | 4.1 |
| 20 | 26 | 8 | 8 | 1664 | 1331 | 3.8 | 2.6 | 1.8 |
| 21 | 30 | 16 | 16 | 7680 | 6144 | 17.4 | 3.0 | 3.6 |
| 22 | 23 | 18.5 | 18.5 | 7872 | 6297 | 17.5 | 2.3 | 4.2 |
| 23 | 120 | 44 | 44 | 232,320 | 185,856 | 527.8 | 12.0 | 10.0 |
| 24 | 28 | 14 | 14 | 5488 | 4390 | 12.5 | 2.8 | 3.2 |
| 25 | 40 | 16 | 16 | 10,240 | 8192 | 23.3 | 4.0 | 3.6 |
| 26 | 18 | 4.5 | 4.5 | 365 | 292 | 0.8 | 1.8 | 1.0 |
| 27 | 26 | 8 | 8 | 1664 | 1331 | 3.8 | 2.6 | 1.8 |
| 28 | 21 | 5 | 5 | 525 | 420 | 1.2 | 2.1 | 1.1 |
| 29 | 19 | 6.5 | 6.5 | 803 | 642 | 1.8 | 1.9 | 1.5 |
| 30 | 17 | 4 | 4 | 272 | 218 | 0.6 | 1.7 | 0.9 |
| Average | 79 | 18 | 15.5 | 28,821 | 22,870 | 64.9 | 4.1 | 3.5 |

**Table A4.** Quantification of clast size, volume, and estimated weight from location 4 on at the west end of Playa El Confital. The density of basalt at 2.84 g/cm$^2$ is applied uniformly in order to calculate wave height for each boulder on the basis of competing equations. Abbreviation: EWH = estimated wave height.

| Sample | Long Axis (cm) | Intermediate Axis (cm) | Short Axis (cm) | Volume (cm$^3$) | Adjustto 80% | Weight (kg) | EWH Nott [17] (m) | EWH Pepe et al. [19] (m) |
|---|---|---|---|---|---|---|---|---|
| 1 | 55 | 38 | 18 | 37,620 | 30,096 | 85.5 | 5.5 | 4.1 |
| 2 | 52.5 | 42 | 21 | 46,305 | 37,044 | 105.2 | 5.3 | 4.8 |
| 3 | 50 | 25 | 22 | 27,500 | 22,000 | 62.48 | 5.0 | 5.0 |
| 4 | 59 | 53.5 | 20 | 63,130 | 50,504 | 143.4 | 5.9 | 4.5 |
| 5 | 35 | 19.5 | 22 | 15,015 | 12,012 | 34.1 | 3.5 | 5.0 |
| 6 | 31 | 21 | 17 | 11,067 | 8854 | 25.1 | 3.1 | 3.8 |
| 7 | 30 | 13 | 6 | 2340 | 1872 | 5.3 | 3.0 | 1.4 |
| 8 | 52 | 20 | 15 | 15,600 | 12,450 | 35.4 | 5.2 | 3.4 |
| 9 | 35 | 31.5 | 15 | 16,538 | 13,230 | 37.6 | 3.5 | 3.4 |
| 10 | 25 | 9 | 8.5 | 1913 | 1530 | 4.3 | 2.5 | 1.9 |
| 11 | 52 | 16 | 8 | 6656 | 5325 | 15.1 | 5.2 | 1.8 |
| 12 | 24 | 23 | 19 | 10,488 | 8390 | 23.8 | 2.4 | 4.3 |
| 13 | 28 | 21 | 7.5 | 4410 | 3528 | 10.0 | 2.8 | 1.7 |
| 14 | 42 | 20 | 10 | 8400 | 6720 | 19.1 | 4.2 | 2.3 |
| 15 | 42 | 29 | 21 | 25,578 | 20,462 | 58.1 | 4.2 | 4.8 |
| 16 | 32 | 19 | 10.5 | 6384 | 5107 | 14.5 | 3.2 | 2.4 |
| 17 | 41 | 37 | 9 | 13,653 | 10,922 | 31.0 | 4.1 | 2.0 |
| 18 | 23 | 12 | 11 | 3036 | 2429 | 6.9 | 2.3 | 2.5 |
| 19 | 38 | 29 | 19 | 20,938 | 16,750 | 47.6 | 3.8 | 4.3 |
| 20 | 22 | 14.5 | 9.5 | 3031 | 2424 | 6.9 | 2.2 | 2.2 |
| 21 | 25 | 9 | 5 | 1125 | 900 | 2.6 | 2.5 | 1.1 |
| 22 | 30 | 11 | 6.5 | 2145 | 1716 | 4.9 | 3.0 | 1.5 |
| 23 | 22 | 11 | 8 | 1936 | 1549 | 4.4 | 2.2 | 1.8 |
| 24 | 49 | 38 | 24 | 44,688 | 35,750 | 101.5 | 4.9 | 5.4 |
| 25 | 51 | 34 | 30 | 52,020 | 41,616 | 147.7 | 5.1 | 6.8 |
| 26 | 41 | 19.5 | 14 | 11,193 | 8954 | 25.4 | 4.1 | 3.2 |
| 27 | 31 | 21 | 7 | 4552 | 3646 | 10.4 | 3.1 | 1.6 |
| 28 | 56.5 | 51 | 20 | 57,630 | 46,104 | 130.9 | 5.7 | 4.5 |
| 29 | 41 | 26.5 | 14 | 15,211 | 12,169 | 34.6 | 4.1 | 3.2 |
| 30 | 55 | 43 | 20 | 25,800 | 20,640 | 58.6 | 4.3 | 4.5 |
| Average | 39 | 25 | 14.5 | 18,530 | 14,823 | 43.1 | 3.9 | 3.3 |

**Table A5.** Quantification of clast size, volume, and estimated weight from location 5 on at the west end of Playa El Confital. The density of basalt at 2.84 g/cm$^2$ is applied uniformly in order to calculate wave height for each boulder on the basis of competing equations. Abbreviation: EWH = estimated wave height.

| Sample | Long Axis (cm) | Intermediate Axis (cm) | Short Axis (cm) | Volume (cm$^3$) | Adjust to 80% | Weight (kg) | EWH Nott [17] (m) | EWH Pepe et al. [19] (m) |
|---|---|---|---|---|---|---|---|---|
| 1 | 57 | 28 | 22 | 35,112 | 28,090 | 79.8 | 5.7 | 5.0 |
| 2 | 76 | 37 | 20 | 56,240 | 44,992 | 127.8 | 7.6 | 4.5 |
| 3 | 38 | 28 | 27 | 28,728 | 22,982 | 65.3 | 3.8 | 6.1 |
| 4 | 46.5 | 30 | 20 | 27,900 | 22,320 | 63.4 | 4.7 | 4.5 |
| 5 | 46 | 34 | 22.5 | 35,190 | 28,152 | 80.0 | 4.6 | 5.1 |
| 6 | 61 | 30 | 21 | 38,430 | 30,744 | 87.3 | 6.1 | 4.8 |
| 7 | 45 | 38 | 11 | 18,810 | 15,048 | 42.7 | 4.5 | 2.5 |
| 8 | 59 | 43 | 38 | 96,406 | 77,125 | 219.0 | 5.9 | 8.6 |
| 9 | 114 | 97 | 27 | 298,566 | 238,853 | 678.3 | 11.9 | 6.1 |
| 10 | 81.5 | 42 | 25 | 85,575 | 68,460 | 194.4 | 8.2 | 5.7 |
| 11 | 43 | 28 | 17.5 | 21,070 | 16,856 | 47.9 | 4.3 | 4.0 |
| 12 | 41 | 28 | 13 | 14,924 | 11,939 | 33.9 | 4.1 | 2.9 |
| 13 | 39 | 33 | 17 | 21,879 | 17,503 | 49.7 | 3.9 | 3.8 |
| 14 | 30 | 25 | 7 | 5250 | 4200 | 11.9 | 3.0 | 1.6 |
| 15 | 22 | 11.5 | 6.5 | 1645 | 1316 | 3.7 | 2.2 | 1.5 |
| 16 | 27.5 | 14.5 | 4 | 1595 | 1276 | 3.6 | 2.8 | 0.9 |
| 17 | 20 | 8 | 5 | 800 | 640 | 1.8 | 2.0 | 1.1 |
| 18 | 32 | 27 | 27 | 23,328 | 18,662 | 53.0 | 3.2 | 6.1 |
| 19 | 19 | 10 | 7.5 | 1425 | 1140 | 3.2 | 1.9 | 1.7 |
| 20 | 17.5 | 16 | 9 | 2520 | 2016 | 5.7 | 1.8 | 2.0 |
| 21 | 27 | 17 | 6.5 | 2984 | 2387 | 6.8 | 2.7 | 1.5 |
| 22 | 22 | 12 | 6 | 1584 | 1267 | 3.6 | 2.2 | 1.4 |
| 23 | 12.5 | 7 | 4 | 350 | 280 | 0.8 | 1.3 | 0.9 |
| 24 | 19 | 11 | 10 | 2090 | 1672 | 4.7 | 1.9 | 2.3 |
| Average | 41.5 | 27 | 15.5 | 34,267 | 27,413 | 77.8 | 4.2 | 3.5 |

**Table A6.** Quantification of clast size, volume, and estimated weight from location 6 on at the west end of Playa El Confital. The density of basal at 2.84 g/cm$^2$ is applied uniformly in order to calculate wave height for each boulder on the basis of competing equations. Abbreviation: EWH = estimated wave height.

| Sample | Long Axis (cm) | Intermediate Axis (cm) | Short Axis (cm) | Volume (cm$^3$) | Adjust to 80% | Weight (kg) | EWH Nott [17] (m) | EWH Pepe et al. [19] (m) |
|---|---|---|---|---|---|---|---|---|
| 1 | 58 | 54 | 34 | 106,488 | 85,190 | 242.0 | 5.8 | 7.7 |
| 2 | 214 | 94 | 50 | 1,005,800 | 804,640 | 2258.0 | 21.5 | 11.3 |
| 3 | 84 | 83 | 71 | 495,012 | 396,010 | 1124.6 | 8.4 | 16.1 |
| 4 | 47 | 33 | 26 | 40,326 | 32,261 | 91.6 | 4.7 | 5.9 |
| 5 | 33 | 17 | 13 | 7293 | 5834 | 16.6 | 3.3 | 2.9 |
| 6 | 50.5 | 49 | 28 | 69,286 | 55,429 | 157.4 | 5.1 | 6.3 |
| 7 | 70.5 | 52 | 32 | 117,312 | 93,850 | 266.5 | 7.1 | 7.2 |
| 8 | 85 | 44 | 32 | 119,680 | 95,744 | 271.9 | 8.5 | 7.2 |
| 9 | 40 | 31 | 13 | 16,120 | 12,896 | 36.6 | 4.0 | 2.9 |
| 10 | 46.5 | 24 | 16 | 17,856 | 14,285 | 40.6 | 4.7 | 3.6 |
| 11 | 34.5 | 27.5 | 17.5 | 16,603 | 13,283 | 37.7 | 3.5 | 4.0 |
| 12 | 31 | 24 | 18 | 13,392 | 10,714 | 30.4 | 3.1 | 4.1 |
| 13 | 40.5 | 24 | 23 | 22,356 | 17,885 | 50.6 | 4.1 | 5.2 |
| 14 | 57 | 41 | 27 | 63,099 | 50,479 | 143.3 | 5.7 | 6.1 |
| 15 | 65 | 60 | 41 | 159,900 | 127,920 | 363.3 | 6.5 | 9.3 |
| 16 | 41 | 38 | 25 | 38,950 | 31,160 | 88.5 | 4.1 | 5.7 |
| 17 | 33 | 21 | 16 | 11,088 | 8870 | 25.2 | 3.3 | 3.6 |
| 18 | 92 | 61 | 26 | 145,912 | 116,730 | 331.5 | 9.2 | 5.9 |
| 19 | 27 | 11 | 11 | 3267 | 2614 | 7.4 | 2.7 | 2.5 |
| 20 | 44 | 41 | 7 | 12,628 | 10,102 | 28.7 | 4.4 | 1.6 |
| 21 | 13 | 9.5 | 5 | 618 | 494 | 1.4 | 1.3 | 1.1 |
| 22 | 30 | 30 | 29 | 26,100 | 20,880 | 59.3 | 3.0 | 6.6 |
| 23 | 41 | 27 | 18 | 19,926 | 15,941 | 45.3 | 4.1 | 4.1 |
| 24 | 41 | 24 | 15 | 14,760 | 11,808 | 33.5 | 4.1 | 3.4 |
| Average | 55 | 38 | 25 | 105,991 | 84,792 | 239.7 | 5.5 | 5.6 |

## References


1. Ávila, S.P.; Johnson, M.E.; Rebelo, A.C.; Baptista, L.; Melo, C.S. Comparison of modern and Pleistocene (MIS 5e) coastal Boulder deposits from Santa Maria Island (Azores Archipelago, NE Atlantic Ocean). *J. Mar. Sci. Eng.* **2020**, *8*, 386. [CrossRef]
2. Johnson, M.E.; Ramalho, R.; da Silva, C.M. Storm-related rhodolith deposits from the Upper Pleistocene and recycled coastal Holocene on Sal island (Cabo Verde Archipelago). *Geosciences* **2020**, *10*, 419. [CrossRef]
3. Scheffers, A.; Kelletat, D. Megaboulder movement by superstorms: A geomorphological approach. *J. Coastal Res.* **2020**, *36*, 844–856. [CrossRef]
4. Johnson, M.E.; Ledesma-Vázquez, J.; Guardado-Grance, R. Coastal geomorphology of a Holocene Hurricane Deposit on a Pleistocene Marine Terrace from Isla Carmen (Baja California Sur, Mexico). *J. Mar. Sci. Eng.* **2018**, *6*, 108. [CrossRef]
5. Johnson, M.E.; Guardado-France, R.; Johnson, E.M.; Ledesma-Vázquez, J. Geomorphology of a Holocene Hurricane deposit eroded from rhyolite sea cliffs on Ensenada Almeja (Baja California Sur, Mexico). *J. Mar. Sci. Eng.* **2019**, *7*, 193. [CrossRef]
6. Johnson, M.E.; Johnson, E.M.; Guardado-France, R.; Ledesma-Vázquez, J. Holocene hurricane deposits eroded as coastal barriers from andesite sea cliffs at Puerto Escondido (Baja California Sur, Mexico). *J. Mar. Sci. Eng.* **2020**, *8*, 75. [CrossRef]
7. van de Bogaard, P. The origin of the Canary Island Seamount Province—New ages of old seamounts. *Sci. Rep.* **2013**, *3*, 2107. [CrossRef]
8. McDougall, I.; Schmincke, H.H. Geochronology of Gran Canaria, Canary Islands: Age of shield building volcanisms and other magmatic phases. *Bull. Volcanol.* **1997**, *40*, 57–77. [CrossRef]
9. Hansen, A. *Los Volcanes Recientes de Gran Canaria*; Rueda-Cabildo Insular de Gran Canaria: Las Palmas, Spania, 1987; 151p.
10. Balcells, R.; Barrera, J.L. Mapa geológico y Memoria de la Hoja 1101-I-II (Las Palmas de Gran Canaria). Mapa Geolópgica de España E. 1:25.000. Seguna Serie (MAGNA), Primera edición. IGME (Institute of Geology and Mineralogy of Spain). 1990; 130p.
11. González-Rodríguez, A.; Melo, C.S.; Galindo, I.; Mangas, J.; Sánchez, N.; Coello, J.; Lozano-Francisco, M.C.; Johnson, M.; Romero, C.; Vegas, J.; et al. Historia geológica y reconstrucción paleobiológica de los depósitos paleontológicas de la Playa de El Confital (Gran Canaria Islas Canarias). In *Yacimientos Paleontológicos Excepcionales en la Península Ibérica*; Cuadernos del Museo Geominero, n° 27; Vaz, N., Sá, A.A., Eds.; Publicaciones del Instituto Geológica y Minero de España: Madrid, Spain, 2018; pp. 491–499. ISBN 978-84-9138-066-5.
12. Bethencourt-González, J.; Dorta-Antequera, P. The storm of November 1826 in the Canary Islands: Possibly a tropical cyclone? *Geogr. Ann. Series A Phys. Geogr.* **2010**, *92*, 329–337. [CrossRef]
13. Wentworth, C.K. A scale of grade and class terms for clastic sediments. *J. Geol.* **1922**, *27*, 377–392. [CrossRef]
14. Ruban, D.A. Costal boulder deposits of the Neogene world: A synopsis. *J. Mar. Sci. Eng.* **2019**, *7*, 446. [CrossRef]

15. Sneed, E.D.; Folk, R.L. Pebbles in the lower Colorado River of Texas: A study in particle morphogenesis. *J. Geol.* **1958**, *66*, 114–150. [CrossRef]
16. Hernández Gutiérrez, L. Caracterización Geomecánica de las Rocas Volcánicass de las Islas Canarias. Ph.D. Thesis, Departamento de Biología Animal, Edafología y Geología, Universidad de la Laguna, Santa Cruz de Tenerife, Spain, 2014; 308p.
17. Nott, J. Waves, coastal bolder deposits and the importance of pre-transport setting. *Earth Planet. Sci. Lett.* **2003**, *210*, 269–276. [CrossRef]
18. Nandasena, N.A.K.; Paris, R.; Tanaka, N. Reassessment of hydrodynamic equations: Minimum flow velocity to initiate boulder transport by high energy events (storms, tsunamis). *Mar. Geol.* **2011**, *281*, 70–84. [CrossRef]
19. Pepe, F.; Corradino, M.; Parrino, N.; Besio, G.; Presti, V.L.; Renda, P.; Calcagnile, L.; Quarta, G.; Sulli, A.; Antonioli, F. Boulder coastal deposits at Favignana Island rocky coast (Sicily, Italy): Litho-structural and hydrodynamic control. *Geomorphology* **2018**, *303*, 191–209.
20. Meco, J.; Guillou, H.; Carracedo, J.C.; Lomoschitz, A.; Ramos, A.J.G.; Rodríguez Yanez, J.J. The maximum warmings of the Pleistocene world climate recorded in the Canary Islands. *Palaeogeogr. Palaeoclimatol. Palaeoecol.* **2002**, *185*, 197–210. [CrossRef]
21. Martín-González, E.; González-Rodríguez, A.; Vera-Peláez, J.L.; Lozano-Francisco, M.C.; Castillo, C. Asociaciones de moluscos de los depósitos litorales del Pleistoceno superior de Tenerife (Islas Canarias, España). *Vieraea* **2016**, *44*, 87–106.
22. Martín-González, E.; González-Rodríguez, A.; Galindo, I.; Mangas, J.; Romero, M.C.; Sánchez, N.; Coello, J.J.; Márquez, A.; Vegas, J.; De Vera, A.; et al. Review of the MIS 5e coastal outcrops from Fuerteventura (Canary islands). *Vieraea* **2019**, *46*, 667–688.
23. Tuya, F.; Ramírez, R.; Sánchez-Jerez, P.; Haroun, R.J.; González-Ramos, A.J.; Coca, J. Coastal resources exploitation can mask bottom up mesoscale regulation of intertidal populations. *Hydrobiologia* **2006**, *553*, 337–344.
24. Haroun, R.J.; Gil-Rogríguez, M.C.; Díaz de Castro, J.; Prud'homme van Reine, W.F. A checklist of the marine plants from the Canary Islands (Central Eastern Atlantic Ocean). *Botanica Marina* **2002**, *45*, 139–169. [CrossRef]
25. Kossin, J.P.; Camargo, S.J.; Sitkowski, M. Climate modulation of North Atlantic hurricane tracks. *J. Climate* **2010**, *23*, 3057–3076. [CrossRef]
26. Borges, P. Ambientes litorais nos grupos Central e Oriental do Arquipélago dos Açores, Conteúdos e Dinamica de Microescale. Unpublished Ph.D. Thesis in Geology, Universidade do Açores, Ponta Delgada, Portugal, 2003; pp. 1–412.
27. Gonçalves, M.; Martinho, P.; Soares, C.G. Assessment of wave energy in the Canary Islands. *Renew. Energy* **2014**, *68*, 774–784. [CrossRef]
28. Gonçalves, M.; Martinho, P.; Soares, C.G. Wave energy assessment based on a 33-year hind-case for the Canary Islands. *Renew. Energy* **2020**, *152*, 258–269. [CrossRef]
29. May, S.M.; Engel, M.; Brill, D.; Squire, P.; Scheffers, A.; Kelletat, D. Coastal Hazards from Tropical Cyclones and Extratropical Winter Storms Based on Holocene Storm Chronologies. In *Coastal World Heritage Sites*; Springer Nature: Berlin, Germany, 2012; Volume 6, pp. 557–585.
30. Johnson, M.E. Holocene boulder beach eroded from chromite and dunite sea cliffs at Støpet on Leka Island (Northern Norway). *J. Mar. Sci. Eng.* **2020**, *8*, 644. [CrossRef]
31. Suárez Rodríguez, C. *Crónicas Urbanas de Historia Natural, Charles Lyell y Pedro Maffiotte en Las Palmas de Gran Canaria*; Idea: Las Palmas, Gran Canaria, 2018; 242p.
32. Déniz González, I. *Patrimonio Geológico Costero en Las Palmas de Gran Canaria: Inventario y Valoración de los Lugares de Interés Geológico*; Acreditación de la Etapa Investigadora (AEI) del Doctorado de Gestión Costera, Universidad de Las Palmas: Las Palmas, Gran Canaria, 2011; 262p.
33. Meco, J.; Koppers, A.A.P.; Miggins, D.P.; Lomoschitz, A.; Betancort, J.F. The Canary record of the evolution of the North Atlantic Pliocene: New 40Ar/39Ar ages and some notable palaeontological evidence. *Palaeogeogr. Palaeoclimatol. Palaeoecol.* **2015**, *435*, 53–69. [CrossRef]
34. Martín-González, E.; Galindo, I.; Vegas, J.; Sánchez, N.; Coello, J.J.; Romero, C.; González-Rodríguez, A. Selección preliminar de Lugares de Interés Paleontológico para el inventario de Patrimonio Geológico de Canarias. *Cuad. Museo Geomin.* **2019**, *30*, 49–55.
35. Déniz González, I.; Mangas, J. Lugares de interés geológico en la costa de Las Palmas de Gran Canaria (Islas Canarias): Inventario y valoración. *Geo-temas* **2012**, *13*, 1253–1256.
36. García-Cortés, A.; Vegas, J.; Carcavilla, L.; Díaz-Martínez, E. *Conceptual Base and Methodology of the Spanish Inventory of Sites of Geological Interest (IELIG)*; Instituto Geológico y Minero de España: Madrid, Spain, 2019; 102p, ISBN 978-84-9138-092-4.
37. Wignall, R.M.L.; Gordon, J.E.; Brazier, V.; MacFadyen, C.C.J.; Everett, N.S. *A Climate Change Risk-Based Assessment for Nationally and Internationally Important Geoheritage Sites in Scotland Including all Earth Science Features in Sites of Special Scientific Interest (SSS)*; Scottish Natural Heritage Research Report No. 1014; Inverness, Scotland, 2018; 53p, Available online: https://www.researchgate.net/profile/Vanessa-Kirkbride/publication/323696506_A_climate_change_risk-based_ assessment_for_nationally_and_internationally_important_geoheritage_sites_in_Scotland_including_all_Earth_science_ features_in_Sites_of_Special_Scientific_Interest_SSSI_Scot/links/5aa66dafa6fdcc29af5308db/A-climate-change-risk-based-assessment-for-nationally-and-internationally-important-geoheritage-sites-in-Scotland-including-all-Earth-science-features-in-Sites-of-Special-Scientific-Interest-SSSI-Scot.pdf (accessed on 28 January 2021).

Journal of
**Marine Science and Engineering**

MDPI

*Article*

# Holocene Boulder Beach Eroded from Chromite and Dunite Sea Cliffs at Støypet on Leka Island (Northern Norway)

**Markes E. Johnson**

Department of Geosciences, Williams College, Williamstown, MA 01267, USA; mjohnson@williams.edu; Tel.: +1-413-2329

Received: 29 June 2020; Accepted: 19 August 2020; Published: 21 August 2020

**Abstract:** This project examines the role of high-latitude storms degrading a Holocene coast formed by igneous rocks composed of low-grade chromite ore and dunite that originated within the Earth's crust near the upper mantle. Such rocks are dense and rarely exposed at the surface by tectonic events in the reconfiguration of old ocean basins. An unconsolidated boulder beach occupies Støypet valley on Leka Island in northern Norway, formerly an open channel 10,000 years ago when glacial ice was in retreat and rebound of the land surface was about to commence. Sea cliffs exposing a stratiform ore body dissected by fractures was subject to wave erosion that shed large cobbles and small boulders into the channel. Competing mathematical equations are applied to estimate the height of storm waves impacting the channel floor and cliffs, and the results are compared with observations on wave heights generated by recent storms striking the Norwegian coast with the intensity of an *orkan* (Norwegian for hurricane). Lateral size variations in beach clasts suggest that Holocene storms struck Leka Island from the southwest with wave heights between 5 and 7.5 m based on the largest beach boulders. This result compares favorably with recent high-latitude storm tracks in the Norwegian Sea and their recorded wave heights. The density of low-grade chromite ore (3.32 g/cm$^3$) sampled from the beach deposit exceeds that of rocks like limestone or other igneous rocks such as rhyolite, andesite, and basalt taken into consideration regarding coastal boulder deposits associated with classic hurricanes in more tropical settings.

**Keywords:** coastal storm deposits; storm surge; hydrodynamic equations; high-latitude settings

---

## 1. Introduction

Global Geoparks authorized by the United Nations Educational, Scientific and Cultural Organization (UNESCO) have expanded in number since inception in 2000 to more than 160 units in 45 different countries. Geology in one form or another is the educational focus of such parks, but with an emphasis on geotourism in support of conservation and the socioeconomic development of rural areas [1]. Remote by nature, many islands possess features of extraordinary significance worthy of development as geoparks. As an example, Iceland is iconic for its status as an island that straddles an expanding oceanic ridge. The island boasts of two UNESCO geoparks, but the potential for several additional parks is anticipated by local planners [2]. In Norway, the Trollfjell (Troll Mountain) Geopark is an example of a well-organized geopark in a remote part of that country near the Arctic Circle that includes many coastal islands with small communities. Stunning coastal scenery that enfolds world-class aspects of geology in the Trollfjell Geopark also includes elements of regional folklore that relate the landscape to its human occupation. Established in 2010, the National Norwegian Geological Monument on Leka Island within the greater geopark boosts the program beyond that envisioned by UNESCO [3]. The monument not only abets an increase in commerce through geotourism but also attracts working geologists and geomorphologists who otherwise may not have known about the place. The dynamics foster a feed-back loop, through which a steady increase in knowledge adds to the overall significance of geoheritage. The present contribution on a unique boulder beach and the interpretation of its hydrodynamics is offered in that spirit.

The aim of this paper is to review unusual physical traits along one of the monument's well-marked trails at Støypet near the north shore of Leka Island. In Norwegian, the word *støypet* may be translated as "the foundry" in reference to the rare igneous rocks and chromite ore concentrated at that place [4,5], although there is no evidence that mineral extraction and smelting took place any time since human occupation began in the earliest Holocene. The unique aspect of Støypet as a geomorphologic and cultural site is the accumulations of a boulder beach dominated by "rolling stones" eroded from adjacent rocky shores composed of dunite and chromite. These ultramafic rocks originated in the deepest part of the earth's crust near the discontinuity with the upper mantle and register high values of specific gravity that give them greater mass. The implication is that wave heights affecting coastal surge were sufficiently powerful to remove joint-bound blocks from sea cliffs that resulted in unusually dense boulders. The same mathematical equations for estimation of wave heights as applied previously to other shores with boulders derived from more common source rocks of lesser mass, including limestone, rhyolite, andesite, and basalt [6–9], are newly applied at Støypet. In addition, the present-day steering winds and wave dynamics characteristic of Norway's Arctic Circle region [10] are reviewed in the context of prominent storms during recent decades in order to appraise the likely direction of wave impact responsible for the boulder beach at Støypet.

## 2. Geographical and Geological Setting

Open to the Norwegian Sea off the coast of northern Norway, Leka Island lies south of the Arctic Circle within the Trollfjell Geopark (Figure 1a). Together with smaller Madsøya, the two islands combine for a total area of 57 km$^2$ (Figure 1b), and support a population of about 500 inhabitants. From the mainland, the island is accessible by ferryboat across a 4-km wide strait. Norway's National Geological Monument offers an extensive system of hiking trails that are well laid out and include educational trail-side markers with texts in Norwegian, English, and German. Several themes combine to make the park an attractive experience, including local folklore. The island's most prominent landmark is a monolith at the side of Lekamøyhammaren Mountain (Figure 1b), said to embody a troll maiden. The monolith (Figure 2a) is a sea stack composed of the igneous rock gabbro, now isolated inland by postglacial rebound at an elevation 100 m above present-day sea level. Other effects of coastal erosion are evident around the sea stack, where former sea cliffs show deep wear in the form of surge channels cut and polished in gabbro basement rocks (Figure 2b). Detailed work by Høgaas and Sveia (2015) that covers the inner (southeastern) part of Leka Island outlines slightly earlier shore erosion by ice scour during the Younger Dryas interval of 12.8 to 11.5 thousand years ago [11].

Geologically, Leka Island is renowned for exposures of the igneous rocks dunite and harzburgite. These are attributed to formation around the boundary between the Earth's lower crust and upper mantle, known as the Mohorovicic Discontinuity [4]. Such rocks are accessible in few other places around the world where their occurrence is justly celebrated, most notably in Canada's Gros Morne National Park in western Newfoundland [12]. Enrichment in minerals belonging to the platinum group, including gold, platinum, and chromium, also occurs as a dense chromite ore associated with dunite on Leka Island [5]. Brought to the surface by tectonic events, these parent rocks were locally subject to Holocene coastal erosion. Thus, today's Leka Island represents a rare convergence of factors related to the geology of deep ocean crust and the geomorphology of recent marine coastal erosion.

On the island's north shore (Figure 1c), access to a Holocene valley crosses a seam of banded chromite framed by outcrops of dunite (Figure 3) that is described by geopark signage. It is noticeable that the chromite bands at this locality are dissected by joints and fractures typical of exfoliation under subaerial conditions comparable to those in former sea cliffs elsewhere in Støypet valley.

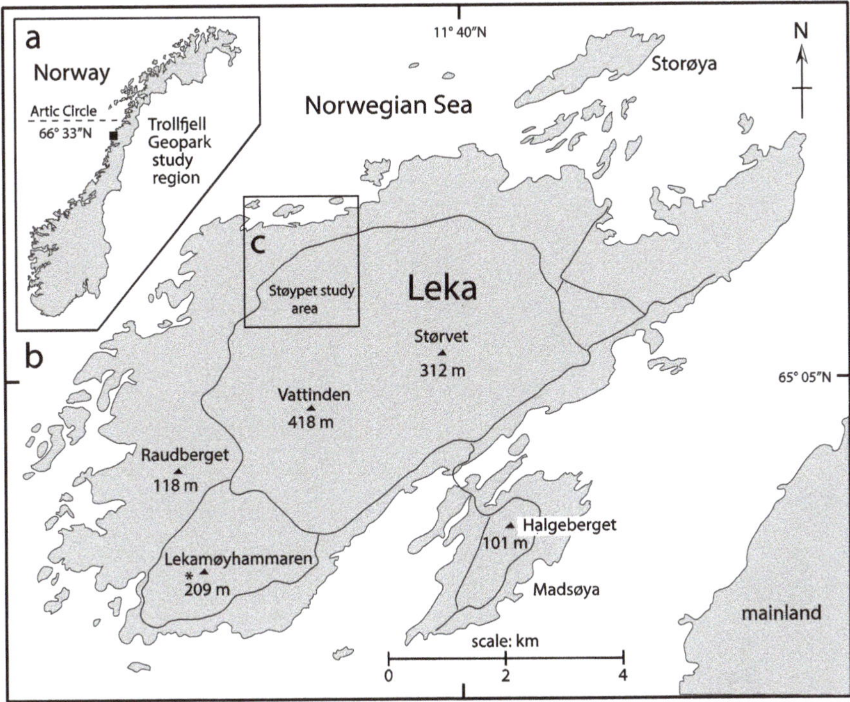



**Figure 1.** The west coast of Norway and Leka Island: (**a**) Norway, showing latitude of the Arctic Circle and region of the Trollfjell Geopark, (**b**) Leka Island and Madsøya with details on highland peaks and location of the Leka maiden monolith (asterisk), and (**c**) inset showing location the Støypet study site.

**Figure 2.** The raised shoreline now about 100 m above sea level at the base of Lekamøyhammaren Mountain (see Figure 1b): (**a**) Gabbro sea stack attributed by legend to the stony embodiment of the Leka troll maiden (author for scale) and (**b**) wave-polished surge channel eroded in gabbro close by the sea stack (figure for scale).

**Figure 3.** Rough exfoliation of chromite bands (dark rocks) and dunite (pale rocks) at the northeast (NE) end of Støypet valley 250 m from the island ring road on the access trail.

## 3. Materials and Methods

### 3.1. Data Collection

Leka Island was visited in July 2019, when the original data for this study were collected from an unconsolidated beach deposit dominated by chromite cobbles and boulders at Støypet. Individual clasts from three stations each limited to collection within a 2-m span were measured manually in three dimensions perpendicular to one another (long, intermediate, and short axes). The stations are confined to a southwest (SW) to northeast (NE) trending valley that contains the Støypet beach deposit. Differentiated from cobbles, the base definition for a boulder adapted in this exercise is that of Wentworth (1922) for an erosional clast equal or greater than 256 mm in diameter [13]. No upper limit for this category is defined in the geological literature [14]. Triangular plots were employed to show variations in clast shape, following the design of Sneed and Folk (1958) for river pebbles [15]. Comparative data on maximum cobble and boulder dimensions were fitted to bar graphs to show size variations in the long and short axes from one sample to the next. Fracture patterns in chromite layers exposed along the valley floor and walls can be examined in two dimensions only, but roughly rectangular outlines can be appraised for a comparison between the original source rocks and range of three-dimensional clast sizes found in the beach deposit. A chromite cobble from the top of Støypet was collected for laboratory determination of density in order to yield weight and volume assigned as a function of equal displacement when submerged in a beaker of water. The commercial "Tour Map" for Leka from 2017 [16] was scanned and adapted in preparation of a detailed base map representing Støypet topography around the study site close to the north shore of Leka Island (see Figure 1c).

### 3.2. Hydraulic Model

Dependent on the calculation of density for low-grade chromite, a hydraulic model may be applied to predict the force needed to remove cobbles and boulders from a rocky shoreline with joint-bound blocks as a function of wave impact. Chromite is an igneous rock that forms in the deepest part of the Earth's crust with variable thicknesses due to zonal banding. These factors mitigate the size and shape of blocks loosened by storm waves once the crustal rocks are brought to the surface. Herein, two formulas were applied to estimate the size of storm waves against joint-bound blocks derived, respectively, from Equation (36) in the work of Nott [17] and from an alternative approach using the velocity equations of Nandasena et al. (2011) [18] as applied by Pepe et al. (2018) [19]:

$$Hs = \frac{\left(\frac{\rho_s - \rho_w}{\rho_w}\right)a}{C_l} \tag{1}$$

$$H_S = \frac{2 \cdot \left(\frac{\rho_s - \rho_w}{\rho_w}\right) \cdot c \cdot (\cos\theta + \mu_s \times \sin\theta)}{C_l}/100 \tag{2}$$

where $Hs$ = height of the storm wave in meters at breaking point, $\rho_s$ = density of the boulder (tons/m$^3$ or g/cm$^3$), $\rho_w$ = density of water at 1.02 g/mL, $a$ = length of boulder on the long axis in cm, $\theta$ is the angle of the bed slope at the pre-transport location (1° for joint-bounded boulders), $\mu_s$ is the coefficient of static friction (= 0.7), $C_l$ is the lift coefficient (= 0.178), and $c$ is length of boulder on the short axis in cm. Equation (1) is sensitive only to the length of a boulder on the long axis, whereas Equation (2) is more sensitive to the length of a boulder on the short axis. Therefore, some differences are expected in the estimates of $H_S$. It is noted that Equation (2) (above) was shown incorrectly in a previous paper dealing with basalt boulder beds from Santa Maria Island in the Azores [9], although the accompanying calculations were performed according to the proper formula.

## 4. Results

### 4.1. Base Map

The base map for the study site at Støypet defines a narrow valley that crosses a topographic saddle between prominent highlands at Steinstind and Hagafjellet, respectively, 190 and 345 m above present-day sea level (Figure 4). The valley is accessed from two endpoints on the Leka Island ring road and follows a well-marked geopark trail for a distance of 2 km. At the topographic saddle between Steinstind and Hagafjellet, the NE to SW trending valley is 50 m wide at an elevation just under 100 m above present sea level (Figure 5a).

The park trail leading from the NE trailhead (Figure 4) climbs a smooth gradient through the deposit to the top located in mid-valley (Figure 5). The view to the northeast across the slope includes the enclosing valley walls with interbedded chromite–dunite layers (Figure 6). Based on horizontal distance in proportion to vertical rise, the slope from the NE direction amounts to 5°. The slope on the opposite side that descends to the SW is similar in vertical drop over horizontal distance, but is broken by a series of cobble-boulder ridges that make it difficult to project a simple gradient.

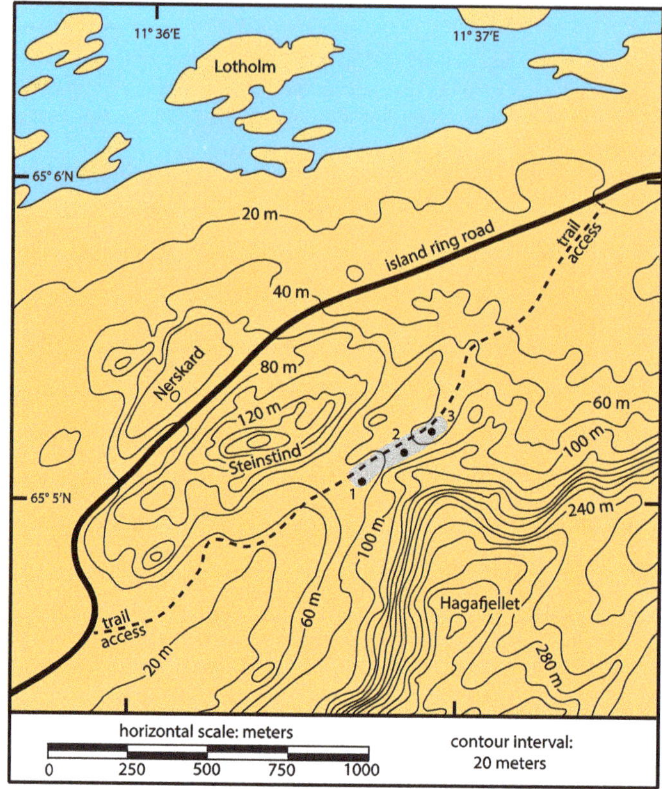



**Figure 4.** Topographic base map for the study area and access trail at Støypet in the National Norwegian Geological Monument on Leka Island. The shaded area near the center represents the limits of an unconsolidated cobble/boulder deposit at the pass between Steinstind and Hagafjellet. Black dots mark the location of three sample sites. See Figure 1c for orientation with respect to the rest of the island. See Figure 1c for location on Leka Island.

**Figure 5.** Details at the top of Støypet within a southwest to northeast (SW-NE) trending valley: (**a**) view to the northwest across the cobble-boulder field perpendicular to the trend of the valley and (**b**) close-up of cobbles and boulders dominated by chromite (darker rocks) with the clast at the center (marked by tape-measure case) having a diameter of 26 cm across the long axis.

**Figure 6.** View from the mid-valley beach deposit over the slope descending to the NE.

*4.2. Fracture Pattern in Støypet Valley Chromite Layers*

The fracture pattern in chromite layers interbedded with dunite is characteristic of basement rocks exposed in the valley floor (Figure 3) and south wall of Støypet valley near its opening to the northeast (Figure 7a). In closer view (Figure 7b), the pattern of vertical joints seen in two dimensions defines columns roughly 40 cm wide. The resulting fractures in combination with horizontal layering results in a two-dimensional outline of rectangular shapes. A set of measurements in two dimensions is represented by a small sample from the outcrop surface. The larger rectangle defined by fractures at the center of Figure 7b is 33 × 21 cm. Smaller rectangles in the layer directly above measure 16 × 9 cm and 11.5 × 10 cm, respectively. Not shown in its entirety (Figure 7b), a much larger rectangle at the side has a height of at least 43 cm. This general pattern would have been subject to hydraulic pressure and plucking during wave impact in the early Holocene when Støypet valley was first flooded. Initially, any blocks so removed from chromite sea cliffs would have been rectilinear in three dimensions prior to the rounding of cobbles and small boulders that resulted from clast abrasion. The size range of cobbles and small boulders in the boulder beach at the top of Støypet (Figure 5a) readily fit with this pattern of fracture-size and shapes. Calculations for 80% of volume from a three-dimensional cubic solution (Appendix A and Tables A1–A3) is a rough estimate for the size reduction of individual cobbles and boulders that underwent abrasion after removal from sea cliffs during contact with one another under storm conditions.

**Figure 7.** Natural fracture pattern in chromite layers interbedded with dunite exposed in the south wall of Støypet valley: (**a**) view to the south at the northeast end of Støypet valley showing chromite layers approximately 4 m thick and (**b**) close-up view showing vertical joints spaced about 40 cm apart.

## 4.3. Chromite Ore Density

The sample of low-grade chromite ore collected at Støypet yielded a value of 3.32 g/cm$^3$ for density based on a small cobble weighing 83 gm and displacing an equivalent volume of water amounting to 25 m. The number is derived by dividing weight by volume. This value was applied uniformly to all clasts listed in Tables A1–A3.

## 4.4. Comparative Variation in Clast Shapes

Raw data on clast size in three dimensions collected from each of the three sample sites are shown in Appendix A (Tables A1–A3). With regard to shape, points representing individual cobbles and boulders are fitted to a set of Sneed–Folk triangular diagrams (Figure 8a–c). The spread of points across these plots consistently shows a strong similarity in the variation of shapes from one sample to another. Few points fall into the upper-most triangle, which represents an origin from a perfectly cube-shaped endpoint as a joint-bound block. The majority of points in each sample fall within the middle part of the two tiers below the top triangle. Those points clustered at the core of any given triangular plot are representative of clasts for which two of the three dimensions are close in value. Relatively few points fall into the middle-right and lower-right domains of the field, which signify a tendency toward development of elongated shapes eroded from source rock exposed in sea cliffs. The composite slope of points suggests a diagonal trend in orientation, although the majority of points cluster vertically within the two central boxes. There is a total absence of points plotting within the left side and lower left part of the plots. The overall aspect of distributed points in these plots signifies the rounding of clasts in which two of the dimensions (preferentially the maximum and intermediate lengths) are closely matched.

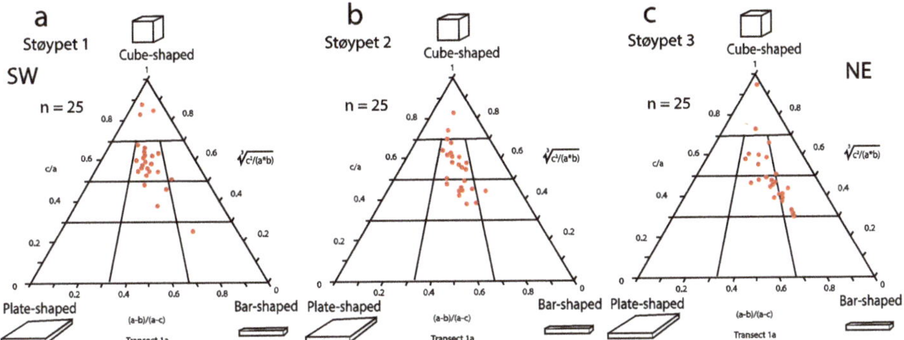

**Figure 8.** Set of triangular Sneed–Folk diagrams used to appraise variations in cobble and boulder shapes: (**a**) trend from sample 1 on the SW flank of Støypet, (**b**) trend sample 2 at the top, and (**c**) Trend from sample 3 on the NE flank of Støypet.

## 4.5. Comparative Variation in Clast Sizes

Clast size is conveniently plotted on bar graphs as a function of maximum and minimum length drawn from the original data (Tables A1–A3). The results for each of three field samples are paired according to size intervals at 5-cm intervals with the boundary between large cobbles and small boulders clearly marked at a diameter above 25 cm (Figure 7). The field locality for sample 1 sits below the crest of the pass between Steinstind and Hagafjellet on the SW side of Støypet valley about 75 m above sea level (Figure 4). The ratio of cobbles to boulders from the sample drawing on maximum clast length is 2:3 (Figure 9a). An equal number of large cobbles and small boulders plot adjacent to one another in bins at the definitional boundary, but the plot is skewed with two individual boulders having a maximum length between 41 and 50 cm. The general shape of clasts from this sample is demonstrated by comparison with the plot for minimum length in which the majority of clasts align below the boundary between cobbles and boulders (Figure 9b). The field locality for sample 2 occurs at the top of the pass between Steinstind and Hagafjellet at the midpoint of Støypet valley just below 100 m in altitude (Figures 4 and 5). Drawing on data for maximum clast length (Table A2), the ratio of large cobbles to small boulders is 1:7 (Figure 9c), which is significantly different from the example in sample 1 (Figure 9a). The number of clasts that plot adjacent to one another in bins at the definitional boundary are sharply divergent, with small boulders outnumbering large cobbles by more than 3 to 1. The graphs for maximum clast length in Figure 9a, c are more alike with respect to outliers of boulders having a maximum length between 41 and 50 cm. With regard to the range in clast sizes drawn from the short axis, sample 2 includes no boulders at all (Figure 9d), which is a departure from sample 1 in this regard (Figure 9b).

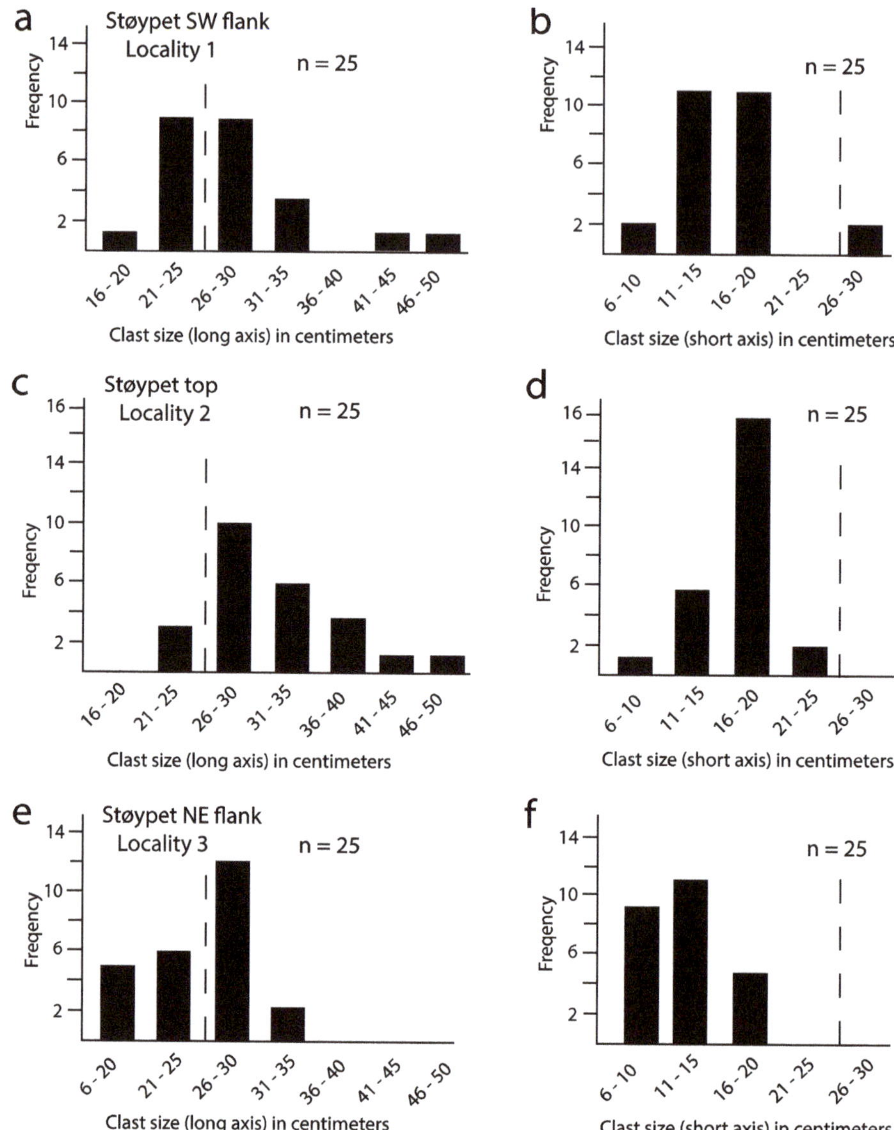


**Figure 9.** Set of bar graphs used to contrast variations in maximum and minimum clast length from three samples in Støypet valley: (**a,b**) bar graphs from Locality 1 on the SW flank of the deposit, (**c,d**) bar graphs from Locality 2 at the top of Støypet, and (**e,f**) bar graphs from Locality 3 on the NE flank of the deposit. Dashed line (offset to represent 26.6 mm) marks the boundary between large cobbles and small boulders.

Like the field locality for sample 1, sample 3 sits at an elevation about 75 m above sea level but on the opposite NE side of Støypet valley below the crest of the pass (Figure 4). Based on data for maximum clast length in this sample (Table A3), the ratio of cobbles to boulders is more balanced, but the quantity of boulders outnumbers that for cobbles and the ratio between large cobbles and boulders in adjacent bins is 1:2 (Figure 9e). There exist fewer outliers of larger boulders in this sample compared to samples 1 and 2 and the number of smaller cobbles in the opposite extreme also exceeds that found in samples 1 and 2 (compared Figure 9a,c,e). Data plotted in bar graphs on the basis of the shortest axis on clasts from sample 3 not only lack boulders but also the class of largest cobbles in size between 21 and 25 cm. In total (Figure 9), the side-by-side plots for sample 3 exhibit a marked downward shift in clast size compared to those from samples 1 and 2.

### 4.6. Storm Intensity as Function of Estimated Wave Height

Average clast sizes and maximum boulder sizes from the three field samples are summarized in Table 1, allowing for direct comparison of mean values for all clasts, as well as values for the largest clasts in each sample based on Equations (1) and (2) derived from the work of Nott [17] and Pepe et al. [19]. The Nott formula [17] shown in Equation (1) yields an average wave height of 3.6 m for the extraction of joint-bound blocks from chromite sea cliffs and their subsequent transfer as beach cobbles and boulders in samples 1–3. According to these results, the mean wave height that impacted shores along the midpoint of Støypet valley and led to the boulder beach on the saddle between Steinstind and Hagafjellet was greatest at 4.1 m. The formula influenced by Pepe et al. [19] showed in Equation (2) yields higher values across all categories with the mean wave height also at the valley's midpoint reaching a value of 4.8 m. Taking into consideration the largest individual boulders from the three samples (Tables A1–A3), Equation (2) differs from Equation (1) in showing a decline in wave heights from the SW end to the NE end of the valley with the most dramatic reduction to 5.2 m (Table 1).

**Table 1.** Summary data from Appendix A (Tables A1–A3) showing maximum bolder size and estimated weight compared to the average values for sampled boulders from each of the transects together with calculated values for wave heights estimated as necessary for boulder-beach mobility. Abbreviations: Max. = maximum, EAWH = estimated average wave height, EMWH = estimated maximum wave height.

| Støypet Locality | Number of Samples | Average Boulder Volume (cm$^3$) | Average Boulder Weight (kg) | EAWH (m) Nott [17] | EAWH (m) Pepe et al. [19] | Max. Boulder Volume (cm$^3$) | Max. Boulder Weight (kg) | EMWH (m) Nott [17] | EMWH (m) Pepe et al. [19] |
|---|---|---|---|---|---|---|---|---|---|
| 1 | 25 | 8136 | 27 | 3.6 | 4.6 | 33,600 | 112 | 6.5 | 7.7 |
| 2 | 25 | 9786 | 33 | 4.1 | 4.8 | 21,120 | 70 | 5.2 | 7.2 |
| 3 | 25 | 4949 | 16 | 3.2 | 3.5 | 10,886 | 36 | 3.6 | 5.2 |
| Average | 25 | 7624 | 25 | 3.6 | 4.3 | 21,869 | 73 | 5.1 | 6.7 |

## 5. Discussion

### 5.1. Physiographic Changes in Island Size

In the context of Leka Island's general physiography, the midpoint of Støypet valley reflects postglacial uplift of the surface from earliest Holocene sea level to nearly 100 m above contemporary sea level mid-valley. This amount of rebound is commensurate with uplift of the sea stack at the SE end of the island said by legend to embody the Leka troll maiden (Figures 1 and 2a). In terms of physical geography prior to uplift, Støypet valley evolved from an open channel that isolated the adjacent heights of Steinstind and Nerskard as a separate entity from the rest of Leka Island. The core of the smaller island would have stood about 80 m above earliest Holocene sea level. The high point of Hagafjellet on the opposite side of the channel was 220 m above earliest Holocene sea level. Cobbles and boulders of chromite and dunite eroded from the facing sea cliffs along the channel began to accumulate as a beach deposit mid-channel, eventually forming a dry connection between the main island and Steinstind. Three archaeological sites are marked on the northwest (NW) embankment overlooking the Støypet channel [16] that were occupied by immigrants who arrived sometime after the local retreat of ice at the end of the Younger Dryas about 11,500 years ago. Cave paintings at nearby Solsem on Leka

Island are dated to the earliest Bronze Age, although Stone Age finds have been excavated from the cave floor [20]. Precise dating for the onset of human occupation in North Trondelag (including Leka Island) is poorly documented, but data from carbon-14 testing in neighboring Nordland and farther north in Troms and Finnmark indicate a pattern of first settlements spanning 9000 to 6000 years ago [21]. Erosion by ice scour along the inner passage between Leka Island and the Norwegian mainland left a distinctive trace correlated with the Main Line now found at a higher elevation between 106 and 112 m above today's sea level [11]. Amalgamation of beach deposits during Holocene time added to the base of the mid-channel connection between Steinstind and Hagafjellet as the land on both sides continued to rebound after ice retreat. The loose "rolling stones" deposited in Støypet valley formed a 50-m wide and 300 m in length plug (Figure 4). In some respects, the deposit on the SW flank may be compared to beach ridges commonly formed in more tropical latitudes as a result of cyclonic storms [22], some of which can be dated by radiometric analysis drawn from coral heads incorporated within distinct ridges. No such method of absolute dating for the testing of storm frequency is possible in Støypet valley, but the centrality of the deposit far from present-day shores at opposite ends of the valley makes it clear that channel erosion ceased long ago.

### 5.2. Direction of Holocene Storms

Application of Equation (1) from Nott [17] and Equation (2) influenced by Pepe et al. [19] differs in results estimating the magnitude of waves impacting sea cliffs along the Holocene Støypet channel, but agrees in the relative ranking of wave energy responsible for transferal of chromite clasts to the three sample sites. Maximum wave energy appears to have focused on the SW part of the channel relative to field sample 1, followed by a reduction in wave energy mid-channel relative to field sample 2, with registration of the lowest wave energy in regard to field sample 3 in the NE part of the channel. From these data (Table 1), it may be argued that initiation of the beach deposit occurred in the vicinity of field sample 1, but that beach ridges pushed farther into the channel as typified by field sample 2. The same amount of surface rebound should have resulted along the entire length of Støypet valley, but the nearly 25 m difference in elevation comparing the altitude of the deposit's axial midpoint (field sample 2) to the its SW and NE extensions is likely due to the formation of a central storm ridge at the same time when uplift began to occur throughout the rest of the island consistent with that around the Leka maiden's Holocene sea stack (Figure 1). The wider mouth of Støypet channel and its emergent valley at one end (Figure 4) also may have influenced the funneling of storm waves that entered from the SW. The smaller clast sizes evident from field sample 3 (Table A3 and Figure 9e,f) can be interpreted as a result of waves that overtopped the central storm ridge and sent lesser clasts down the apparent lee side of the beach to accumulate on a smooth 5° slope. Such a hypothesis posits that early Holocene storms in the Norwegian Sea were more likely to have arrived from the SW, trending to the NE against the adjacent Norwegian mainland. The downward shift in clast sizes from sample 3 (Figure 9e,f) compared to samples 1 (Figure 9a,b) and 2 (Figure 9c,d) implies that a different wave regime may have been in place on the NE end of the valley compared to the SW end.

### 5.3. Inference from Historical Storms

Observations on the steering winds in Norway's Arctic Vestifjord district north of Leka Island are summarized by Jones et al. (1997), based on information derived from weather stations on small islands in the Norwegian Sea as well as the Norwegian mainland [10]. Cod fishing in this district has a long history as a major industry dating back centuries [23], and the difference between winter and summer weather is well known. During the winter months when the fishing season is in play, the dominant winds arrive from the southwest commonly interpreted as fresh breezes between 4 and 5 on the Beaufort scale. However, there is a 10% chance that gale-force winds (8 on the Beaufort scale) will occur with wind speeds reach 20.7 m/s. During the summer months, lighter winds generally arrive from the NE and gale-force winds are rare, comprising less than 1% of station observations [10]. Hurricanes are a seasonal feature of tropical and subtropical settings uncharacteristic for Boreal Seas.

Technically, they are a tropical to subtropical phenomenon that depends on high ocean-surface water temperatures and excessive air moisture [22]. During the last decade (2011 to 2020), four storms assigned to the category of "orkan" by the Norwegian Meteorological Institute [24–27] struck the Norwegian coast including the vicinity of Leka Island (Table 2). These specific storms share a general history of duration lasting 48 h, or more, with a frequency of arrival every second year. Clearly, these factors play a role in beach dynamics related to clast size.

**Table 2.** Impact of superstorms in the North Trondelag and Nordland districts of coastal Norway summarized from reports issued by the Norwegian Meteorological Institute.

| Storm Name | Date of Landfall | Direction of Arrival | Maximum Wind Speed (m/s) | Maximum Wave Height (m) |
| --- | --- | --- | --- | --- |
| Dagmar [24] | 25 December 2011 | SW to W | 35–40 | Unknown |
| Hilde [25] | 16 November 2013 | SW | 40–50 | 12–24 |
| Tor [26] | 7 January 2016 | SW | 35–45 | 9–12 |
| Cora [27] | 1 January 2018 | SW to NW | 45–55 | Unknown |

The direct translation of the Norwegian word "orkan" to English is hurricane, and the NMI's basis for such storms is defined by a minimum wind speed of 32.7 m/s. Major storms in high latitudes express a cyclonic circulation similar to hurricanes, but originate independently of water vapor arising from excessively high ocean-surface water temperatures. Instead, they rely on the acceleration of weather fronts with extreme contrasts in air temperature on opposite sides of the line [28]. Based on the Saffir–Simpson hurricane wind scale, a wind speed of 32.7 m/s falls below the range of a Category 1 hurricane with wind speeds between 33 and 42 m/s. During the last decade, at least two major storms reached the North Trondelag and Nordland districts of Norway packing maximum wind speeds of 50 m/s or higher that according to the Saffir–Simpson scale qualify as Category 3 disturbances. Based on standard energy calculations in mega joules, a high-latitude storm of the kind reaching the mid-section of Norway in recent years expends roughly 50% of the energy of a large tropical hurricane [28], but such a release is sufficient to do extensive damage to coastal infrastructure and erode coastal shores. It is notable that all four superstorms reaching North Trondelag and Nordland including Leka Island arrived from the southwest (Table 2). The local climate in this area following the retreat of glaciers 10,000 years ago was sure to have been more extreme than today, but the arrival pattern of today's major storms fits with the physical characteristics of Holocene boulder deposits filling Støypet valley. Moreover, it is notable that even the lower range of storm-wave heights reported for Hilde and Tor [25,26] exceed the storm-wave heights estimated to have entered the Holocene channel.5.4. Contrast with Coastal Deposits Elsewhere

Holocene deposits formed by unconsolidated cobbles and boulders are widely distributed all around the world, but studies in coastal geomorphology seldom take into account rock density as related to variability in parent rock types when investigating the range of wave heights necessary for their development. Equation (1) as derived from Nott [17] has been applied to coastal boulder deposits throughout Mexico's Gulf of California, including those formed by limestone, rhyolite, and andesite clasts [6–8]. Extension of this work to include Equation (2) as influenced by Pepe et al. [19] also has been applied to coastal basalt deposits in the Azores [9]. The variation in density among these rock types ranges from 1.86 g/cm$^3$ for limestone, 2.16 g/cm$^3$ for the rhyolite, 2.55 g/cm$^3$ for andesite, and 3.0 g/cm$^3$ for basalt. Cobbles and boulders of low-grade chromite ore that are the subject of the present study register a higher density measured as 3.32 g/cm$^3$ and high-grade chromite ore is known to yield an even greater value. Limestone differs from the others as a marine product derived mostly from organic materials, whereas the rest are igneous rocks some of which like rhyolite and andesite typically form under subaerial conditions as volcanic flows. The limestone, rhyolite, and andesite all produce stratified bodies that sooner or later become subject to joints that break perpendicular to the bedding plane. Basalt may issue under subaerial conditions subject to jointing in the same way, but also forms under submarine conditions that entail a different style of accumulation as pillow-shaped bodies, typical on Santa Maria Island in the Azores [9]. Dunite and chromite as found exposed on Norway's Leka Island are igneous rocks that originated deep within the Earth's crust but also under conditions that allowed for stratified bodies associated with serial injections of magma that cooled slowly one after the other. The key similarity among all these rock types is the appearance of horizontal partings cut by joints and fractures. In sea cliffs subject to wave erosion, it is the configuration of layering dissected by joints common to all such rocks that determines the effectiveness of storms to detach blocks subsequently incorporated in coastal boulder deposits.

Rock density is an important factor in coastal erosion, because a storm wave of any given height will behave differently depending on the degree of stratification and jointing in the parent sea cliff. The same wave will have the capacity to dislodge a larger, less dense block of limestone compared to a smaller, denser block of basalt or chromite. Another factor to be considered is the difference between tropical hurricanes limited to lower latitudes and Boreal storms characteristic of higher, more poleward latitudes. The largest block detached from a limestone sea cliff by Holocene hurricanes in the Gulf of California is estimated to weigh 28 metric tons [6], whereas limestone blocks between 100 and 200 metric tons are attributed to detachment from sea cliffs in the Philippines during Super Typhoon Haiyan in 2013 [29]. However, the erosional effectiveness of tropical hurricanes compared to Boreal storms is not to be underestimated on account of movements in blocks weighing as much as 620 metric tons during the winter storms of 2013–2014 against limestone sea cliffs in western Ireland [30]. By comparison, the chromite cobbles and boulders entrained as a Holocene beach deposit in Støypet valley are exceedingly small, but with a much greater density than limestone. Studies on the geomorphology of boulder deposits composed of other rock types such as granite and gabbro are to be encouraged.

## 6. Conclusions

Study of the cobble-boulder beach entrained in Støypet valley on Norway's Leka Island offers insights based on mathematical equations for estimation of Holocene wave heights and wave heights from recent superstorms in the same region:

- The unconsolidated cobble-boulder beach preserved as a Holocene deposit in Støypet valley is unique due to components of low-grade chromite ore and the igneous rock dunite that originated deep within the Earth's crust. In particular, chromite is a dense rock seldom exposed at the surface and more rarely in coastal settings subject to wave erosion.
- Present-day Støypet valley originated as a marine channel that separated part of the island's north shore from the rest of the island in early Holocene time when free passage from one end to the other end was possible prior to surface rebound after the retreat of glacial ice. The valley's mid-point is close to 100 m above present-day sea level with the upper 25 m occupied by the cobble-boulder beach.
- Data on size variations in clast size from three field samples in Støypet valley suggest that Holocene storms entered the former channel from the SW, which is consistent with the approach by recent storms of hurricane strength on the same coast in North Trondelag and Nordland near the Polar Circle.
- The density of constituent clasts derived from low-grade chromite ore eroded by storm waves from joint-bound sea cliffs is 45% more dense than limestone boulders and 25% more dense than andesite boulders previously studied in coastal deposits elsewhere. Although chromite boulders from Leka Island are small in comparison, the energy expended by storm waves to free blocks of chromite from joint-bound sea cliffs would have been greater normalized for unit volume than for limestone or other igneous rocks like andesite.
- Different equations used to estimate the height of storm waves eroding sea cliffs with joint-bound rocks differ in results as found in the formulations applied from Nott [17] and updated by

Pepe et al. [19]. In this study, the latter yielded estimates from 12% to 24% higher depending on analysis of mean clast size or maximum clast size. Holocene wave heights estimated by both equations are within the range of observed wave heights during major storms in the same region.

The Norwegian National Geological Monument within the Trollfjell (Troll Mountain) Geopark already provides an outstanding resource for visitors of all ages and educational backgrounds to learn about earth processes and achieve a better appreciation for our common geoheritage. Its significance can be expected to grow with visitors offering fresh input from different perspectives.

**Funding:** The research project received no external funding.

**Acknowledgments:** B. Gudveig Baarli organized the logistics for a visit to Leka Island and assisted in the collection of field data in Støypet valley in July 2019. Sérgio P. Ávila (University of the Azores) provided the calculations for wave heights based on the mathematical model influenced by Pepe et al. (2018). The author is grateful for recommendations offered by three anonymous reviewers that helped him to improve the final product, as well as crucial comments provided by the scientific advisor.

**Conflicts of Interest:** The author declares no conflict of interest.

## Appendix A

**Table A1.** Quantification of clast size, volume, and estimated weight from a sample on the SW flank of Støypet (locality 1). The density of low-grade chromite at 3.32 g/cm$^3$ is applied uniformly in order to calculate wave height for each boulder. Abbreviation: EWH = estimated wave height.

| Sample | Long Axis (cm) | Intermediate Axis (cm) | Short Axis (cm) | Volume (cm$^3$) | Adjust to 80% | Weight (kg) | Ewh Nott [17] (m) | Ewh Pepe et al. [19] (m) |
|---|---|---|---|---|---|---|---|---|
| 1 | 29 | 19 | 16 | 8816 | 7053 | 23 | 3.7 | 4.6 |
| 2 | 22 | 18 | 18 | 7128 | 5702 | 19 | 2.8 | 5.2 |
| 3 | 24 | 22 | 12 | 6336 | 5069 | 17 | 3.1 | 3.4 |
| 4 | 23 | 14 | 12 | 3864 | 3091 | 10 | 3.0 | 3.4 |
| 5 | 25 | 18 | 6 | 2700 | 2160 | 7 | 3.2 | 1.7 |
| 6 | 18 | 14 | 11 | 2772 | 2218 | 7 | 2.3 | 3.1 |
| 7 | 32 | 17 | 12 | 6528 | 5222 | 17 | 4.1 | 3.4 |
| 8 | 23 | 20 | 20 | 9200 | 7360 | 24 | 3.0 | 5.7 |
| 9 | 32 | 17 | 15 | 8160 | 6528 | 22 | 4.1 | 4.3 |
| 10 | 34 | 20 | 20 | 13,600 | 10,880 | 36 | 4.4 | 5.7 |
| 11 | 30 | 20 | 20 | 12,000 | 9600 | 32 | 3.9 | 5.7 |
| 12 | 26 | 19 | 15 | 7410 | 5928 | 20 | 3.4 | 4.3 |
| 13 | 50 | 30 | 28 | 42,000 | 33,600 | 112 | 6.5 | 8.0 |
| 14 | 34 | 23 | 20 | 15,640 | 12,512 | 42 | 4.4 | 5.7 |
| 15 | 27 | 24 | 17 | 11,016 | 8813 | 29 | 3.5 | 4.9 |
| 16 | 24 | 24 | 20 | 11,520 | 9216 | 31 | 3.1 | 5.7 |
| 17 | 22 | 17 | 10 | 3740 | 2992 | 10 | 2.8 | 2.9 |
| 18 | 28 | 20 | 18 | 10,080 | 8064 | 27 | 3.6 | 5.2 |
| 19 | 26 | 18 | 16 | 7488 | 5990 | 20 | 3.4 | 4.6 |
| 20 | 23 | 17 | 15 | 5865 | 4692 | 16 | 3.0 | 4.3 |
| 21 | 44 | 30 | 27 | 35,640 | 28,512 | 95 | 5.7 | 7.7 |
| 22 | 26 | 20 | 14 | 7280 | 5824 | 19 | 3.4 | 4.0 |
| 23 | 26 | 17 | 14 | 6188 | 4950 | 16 | 3.4 | 4.0 |
| 24 | 26 | 14 | 14 | 5096 | 4077 | 14 | 3.4 | 4.0 |
| 25 | 23 | 14 | 13 | 4186 | 3349 | 11 | 3.0 | 3.7 |
| Average | 27.9 | 19.4 | 16.1 | 10,170 | 8136 | 27 | 3.6 | 4.6 |

**Table A2.** Quantification of clast size, volume, and estimated weight from a sample at the top of the pass at Støypet (Locality 2). The density of low-grade chromite at 3.32 g/cm$^3$ is applied uniformly in order to calculate wave height for each boulder. Abbreviation: EWH = estimated wave height.

| Sample | Long Axis (cm) | Intermediate Axis (cm) | Short Axis (cm) | Volume (cm$^3$) | Adjust to 80% | Weight (kg) | EWH Nott [17] (m) | EWH Pepe et al. [19] (m) |
|---|---|---|---|---|---|---|---|---|
| 1 | 38 | 21 | 16 | 12,768 | 10,214 | 34 | 4.9 | 4.6 |
| 2 | 33 | 22 | 19 | 13,794 | 11,035 | 37 | 4.3 | 5.4 |
| 3 | 42 | 26 | 19 | 20,748 | 16,995 | 56 | 5.4 | 5.4 |
| 4 | 27 | 20 | 12 | 6480 | 5184 | 17 | 3.3 | 3.4 |
| 5 | 32 | 16 | 15 | 7680 | 6144 | 20 | 4.1 | 4.3 |
| 6 | 27 | 15 | 10 | 4050 | 3240 | 11 | 3.3 | 2.9 |
| 7 | 22 | 20 | 18 | 7920 | 6336 | 21 | 2.8 | 5.2 |
| 8 | 33 | 23 | 16 | 12,144 | 9715 | 32 | 4.3 | 4.6 |
| 9 | 26 | 19 | 18 | 8892 | 7113 | 24 | 3.4 | 5.2 |
| 10 | 36 | 22 | 16 | 12,672 | 10,137 | 34 | 4.7 | 4.6 |
| 11 | 24 | 19 | 13 | 5928 | 4742 | 16 | 3.1 | 3.7 |
| 12 | 29 | 20 | 18 | 10,440 | 8352 | 28 | 3.7 | 5.2 |
| 13 | 26 | 20 | 19 | 9880 | 7904 | 26 | 3.4 | 5.4 |
| 14 | 26 | 17 | 16 | 7072 | 5658 | 19 | 3.4 | 4.6 |
| 15 | 46 | 26 | 20 | 23,920 | 19,136 | 64 | 5.9 | 5.7 |
| 16 | 29 | 20 | 11 | 6380 | 5104 | 17 | 3.7 | 3.1 |
| 17 | 37 | 28 | 25 | 23,125 | 18,500 | 61 | 4.8 | 7.2 |
| 18 | 30 | 29 | 13 | 11,310 | 9048 | 30 | 3.9 | 3.7 |
| 19 | 35 | 29 | 20 | 20,300 | 16,240 | 54 | 4.5 | 5.7 |
| 20 | 33 | 23 | 20 | 15,180 | 12,144 | 40 | 4.3 | 5.7 |
| 21 | 30 | 19 | 19 | 10,830 | 8664 | 29 | 3.9 | 5.4 |
| 22 | 23 | 18 | 14 | 5796 | 4637 | 15 | 3.0 | 4.0 |
| 23 | 30 | 22 | 17 | 11,220 | 8976 | 30 | 3.9 | 4.9 |
| 24 | 40 | 30 | 22 | 26,400 | 21,120 | 70 | 5.2 | 6.3 |
| 25 | 34 | 18 | 17 | 10,404 | 8323 | 28 | 4.4 | 4.9 |
| Average | 31.5 | 21.7 | 17 | 12,213 | 9786 | 33 | 4.1 | 4.8 |

**Table A3.** Quantification of clast size, volume, and estimated weight from a sample on the NE flank of Støypet (Locality 3). The density of low-grade chromite at 3.32 g/cm$^3$ is applied uniformly in order to calculate wave height for each boulder. Abbreviation: EWH = estimated wave height.

| Sample | Long Axis (cm) | Intermediate Axis (cm) | Short Axis (cm) | Volume (cm$^3$) | Adjust to 80% | Weight (kg) | EWH Nott [17] (m) | EWH Pepe et al. [19] (m) |
|---|---|---|---|---|---|---|---|---|
| 1 | 29 | 20 | 16 | 9280 | 7424 | 25 | 3.7 | 4.6 |
| 2 | 30 | 19 | 18 | 10,260 | 8208 | 27 | 3.9 | 5.2 |
| 3 | 34 | 20 | 16 | 10,880 | 8704 | 29 | 4.4 | 4.6 |
| 4 | 26 | 15 | 15 | 5850 | 4680 | 16 | 3.4 | 4.3 |
| 5 | 29 | 20 | 14 | 8120 | 6496 | 22 | 3.7 | 4.0 |
| 6 | 26 | 21 | 12 | 6552 | 5242 | 17 | 3.4 | 3.4 |
| 7 | 24 | 20 | 14 | 6720 | 5376 | 18 | 3.1 | 4.0 |
| 8 | 21 | 16 | 8 | 2688 | 2150 | 7 | 2.7 | 2.3 |
| 9 | 20 | 19 | 13 | 4940 | 3952 | 13 | 2.6 | 3.7 |
| 10 | 17 | 17 | 16 | 4624 | 3699 | 12 | 2.2 | 4.6 |
| 11 | 18 | 15 | 13 | 3510 | 2808 | 9 | 2.3 | 3.7 |
| 12 | 26 | 17 | 10 | 4420 | 3536 | 12 | 3.4 | 2.9 |
| 13 | 27 | 20 | 16 | 8640 | 6912 | 23 | 3.5 | 4.6 |
| 14 | 27 | 20 | 10 | 5400 | 4320 | 14 | 3.5 | 2.9 |
| 15 | 32 | 24 | 15 | 11,520 | 9216 | 31 | 4.1 | 4.3 |

**Table A3.** *Cont.*

| Sample | Long Axis (cm) | Intermediate Axis (cm) | Short Axis (cm) | Volume (cm$^3$) | Adjust to 80% | Weight (kg) | EWH Nott [17] (m) | EWH Pepe et al. [19] (m) |
|---|---|---|---|---|---|---|---|---|
| 16 | 24 | 12 | 11 | 3168 | 2534 | 8 | 3.1 | 3.1 |
| 17 | 30 | 20 | 13 | 7800 | 6240 | 21 | 3.9 | 3.7 |
| 18 | 28 | 27 | 12 | 13,608 | 10,886 | 36 | 3.6 | 3.4 |
| 19 | 20 | 15 | 6 | 1800 | 1440 | 5 | 2.6 | 1.7 |
| 20 | 18 | 13 | 8 | 1872 | 1498 | 5 | 2.3 | 2.3 |
| 21 | 26 | 20 | 8 | 4160 | 3328 | 11 | 3.4 | 2.3 |
| 22 | 26 | 22 | 13 | 7436 | 5949 | 20 | 3.4 | 3.7 |
| 23 | 25 | 20 | 8 | 4000 | 3200 | 11 | 3.2 | 2.3 |
| 24 | 25 | 20 | 10 | 5000 | 4000 | 13 | 3.2 | 2.9 |
| 25 | 20 | 15 | 8 | 2400 | 1920 | 6 | 2.6 | 2.3 |
| Average | 25 | 18.7 | 12 | 6186 | 4949 | 16 | 3.2 | 3.5 |

## References


1. Farsani, N.T.; Coelho, C.; Costa, C. Geotourism and geoparks as novel strategies for socio-economic development in rural areas. *Int. J. Tour. Res.* **2011**, *13*, 68–81. [CrossRef]
2. Olafsdóttir, R.; Dowling, R. Geotourism and geoparks—A tool for geoconservation and rural development in vulnerable environments: A case study from Iceland. *Geoheritage* **2014**, *6*, 71–87. [CrossRef]
3. Dahl, R.M.E.; Carstens, H.; Haukdal, G. The election of a National Norwegian Geological Monument: At tool for raising awareness of geological Heritage. *GeoJ. Tour. Geosites* **2011**, *4*, 178–184.
4. Prestvik, T. Alpine-type mafic and ultramafic rocks of Leka, Nord-Trøndelog. *Nor. Geol. Unders.* **1972**, *273*, 23–24.
5. Pedersen, R.-B.; Johannesen, G.M.; Boyd, R. Stratiform platinum-group element mineralization in the ultramafic cumulates of the Leka Ophiolite Complex, Central Norway. *Econ. Geol.* **1993**, *88*, 782–803. [CrossRef]
6. Johnson, M.E.; Ledesma-Vázquez, J.; Guardado-France, R. Coastal geomorphology of a Holocene hurricane deposit on a Pleistocene marine terrace from Isla Carmen (Baja California Sur, Mexico). *J. Mar. Sci. Eng.* **2018**, *6*, 108. [CrossRef]
7. Johnson, M.E.; Guardado-France, R.; Johnson, E.M.; Ledesma-Vázquez, J. Geomorphology of a Holocene Hurricane deposit eroded from rhyolite sea cliffs on Ensenada Almeja (Baja California Sur, Mexico). *J. Mar. Sci. Eng.* **2019**, *7*, 193. [CrossRef]
8. Johnson, M.E.; Guardado-France, R.; Ledesma-Vázquez, J. Holocene hurricane deposits eroded as coastal barriers from andesite sea cliffs at Puerto Escondido (Baja California Sur, Mexico). *J. Mar. Sci. Eng.* **2020**, *8*, 75. [CrossRef]
9. Ávila, S.P.; Johnson, M.E.; Rebelo, A.C.; Baptista, L.; Melo, C.S. Comparison of modern and Pleistocene (MIS 5e) coastal Boulder deposits from Santa Maria Island (Azores Archipelago, NE Atlantic Ocean). *J. Mar. Sci. Eng.* **2020**, *8*, 386. [CrossRef]
10. Jones, B.; Boudjelas, S.; Mitchelson-Jacob, E.G. Topographic steering of winds in Vestfjorden, Norway. *Weather* **1997**, *52*, 304–311. [CrossRef]
11. Høgaas, F.; Sveian, H. The Younger Dryas Main Line on Leka Norway, as determined from a high resolution digital elevation model derived from airborne LiDAR data. *Geomorphology* **2015**, *231*, 63–71. [CrossRef]
12. Burzynski, M. *Gros Morne National Park*; Breakwater Books, Ltd.: St. John's, NL, Canada, 1999.
13. Wentworth, C.K. A scale of grade and class terms for clastic sediments. *J. Geol.* **1922**, *27*, 377–392. [CrossRef]
14. Ruban, D.A. Costal boulder deposits of the Neogene world: A synopsis. *J. Mar. Sci. Eng.* **2019**, *7*, 446. [CrossRef]
15. Sneed, E.D.; Folk, R.L. Pebbles in the lower Colorado River of Texas: A study in particle morphogenesis. *J. Geol.* **1958**, *66*, 114–150. [CrossRef]
16. Leka Commune Authority. *Leka Turkart (målestokk 1:35 000). Single-Sheet Folded Map*; Leka Commune Authority: Stockholm, Sweden, 2017.
17. Nott, J. Waves, coastal bolder deposits and the importance of pre-transport setting. *Earth Planet. Sci. Lett.* **2003**, *210*, 269–276. [CrossRef]

18. Nandasena, N.A.K.; Paris, R.; Tanaka, N. Reassessment of hydrodynamic equations: Minimum flow velocity to initiate boulder transport by high energy events (storms, tsunamis). *Mar. Geol.* **2011**, *281*, 70–84. [CrossRef]

19. Pepe, F.; Corradino, M.; Parrino, N.; Besio, G.; Presti, V.L.; Renda, P.; Calcagnile, L.; Quarta, G.; Sulli, A.; Antonioli, F. Boulder coastal deposits at Favignana Island rocky coast (Sicily, Italy): Litho-structural and hydrodynamic control. *Geomorphology* **2018**, *303*, 191–209. [CrossRef]

20. Nash, G.; Smiseth, M.T. Art and intimacy within the prehistoric landscapes of Norway. In *Ritual Landscapes and Borders within Rock Art Research*; Stebergløkken, H., Berge, R., Lindgaard, E., Stuedal, H.V., Eds.; Archaeopress Publ. Ltd.: Oxford, UK, 2015; pp. 31–46. ISBN 978-1-78491-159-1.

21. Møller, J.J. Shoreline relation and prehistoric settlement in northern Norway. *Nor. Geogr. Tidsskr.* **1987**, *41*, 45–60. [CrossRef]

22. May, S.M.; Engel, M.; Brill, D.; Squire, P.; Scheffers, A.; Kelletat, D. Coastal hazards from tropical cyclones and extratropical winter storms based on Holocene storm chronologies. In *Coastal Hazards*; Finkl, C.W., Ed.; Springer: Berlin/Heidelberg, Germany, 2013; Volume 6, pp. 557–585.

23. Dass, P. *The Trumpet of Nordland*; Helgaland Museum: Mosjøen, Norway, 2015.

24. Kvamme, D.; Abildsnes, H. Ekstremev Ær Rapport DAGMAR, 25 December 2011. Norsk Meteorologisk Institute. 2012, Dagmar E-Rapport. p. 52. Available online: https://www.met.no/publikasjoner/met-info/ekstremvaer/_/attachment/download/69803fbb-7c26-41ce-83e9-0479891c043d:c28f40bfb1fd2c4a07672bd8132b40bcc44717c3/dagmar-e-rapport.pdf (accessed on 21 August 2020).

25. Anonymous. Ekstremev Ær Rapport HILDE, 16–17 November 2013. Norsk Meteorologisk Institute. 2013. Hilde E-Rapport. p. 19. Available online: https://www.met.no/publikasjoner/met-info/ekstremvaer/_/attachment/download/eacf68c3-9237-417b-b8bc-6feb2af01dea:ccc0476b8d4ecb716bd14a80b8da590c71971b55/ekstrem_rapport_hilde.pdf (accessed on 21 August 2020).

26. Anonymous. Ekstremev Ær Rapport TOR, 29–30 January 2016. Norsk Meteorologisk Institute. 2016. Tor E-Rapport. p. 19. Available online: https://www.met.no/publikasjoner/met-info/ekstremvaer/_/attachment/download/f9f30fbc-e10a-43bf-8ac8-54f35139ced7:04f39c24d58c79b43cdcf924b62e4192e5d3926c/MET-info-14-2016.pdf (accessed on 21 August 2020).

27. Olsen, A.-M.; Berge, H. Ekstremev Ær Rapport CORA, January 2018. Norsk Meteorologisk Institute. 2018. Cora E-Rapport. p. 16. Available online: https://www.met.no/publikasjoner/met-info/ekstremvaer/_/attachment/download/099ee978-df9a-4a35-88f1-9b8ece0ccea1:d9f7f511ab78b9e517275c3818b72880d4626310/Met-info-14-2018.pdf (accessed on 21 August 2020).

28. Pielou, E.C. *The Energy of Nature*; University of Chicago Press: Chicago, IL, USA, 2001.

29. Kennedy, A.B.; Mori, N.; Yasuda, T.; Shimozono, T.; Tomiczek, T.; Donahue, A.; Shimura, T.; Imai, Y. Extreme block and boulder transport along a cliffed coastline (Calicoan Island, Philippines) during Super Typhoon Haiyan. *Mar. Geol.* **2017**, *383*, 65–77. [CrossRef]

30. Cox, R.; Jahn, K.L.; Watkins, O.G.; Cox, P. Extraordinary boulder transport by storm waves (west of Ireland, winter 2013–2014), and criteria for analyzing coastal boulder deposits. *Earth Sci. Rev.* **2018**, *177*, 623–636. [CrossRef]

Journal of
*Marine Science
and Engineering*

*Article*

# Finding Coastal Megaclast Deposits: A Virtual Perspective

Dmitry A. Ruban

K.G. Razumovsky Moscow State University of Technologies and Management (the First Cossack University), 109004 Moscow, Russia; ruban-d@mail.ru

Received: 7 February 2020; Accepted: 25 February 2020; Published: 3 March 2020

**Abstract:** Coastal megaclast deposits are dominated by detrital particles larger than 1 m in size. These attract significant attention of modern researchers because of the needs of sedimentary rock nomenclature development and interpretation of storm and tsunami signatures on seashores. If so, finding localities that exhibit coastal megaclast deposits is an important task. Field studies do not offer a quick solution, and, thus, remote sensing tools have to be addressed. The application of the Google Earth Engine has permitted to find four new localities, namely Hondarribia in northern Spain (Biscay Bay), the Ponza Island in Italy (Tyrrhenian Sea), the Wetar Island in eastern Indonesia (Banda Sea), and the Humboldt o Coredo Bay at the Colombia/Panama border (eastern Pacific). In these localities, coastal megaclast deposits consisting of blocks (1–10 m in size) and some megablocks (>10 m in size) are delineated and preliminary described in regard to the dominant size of particles, package density, mode of occurrence, etc. The limitations of such virtual surveys of coastal megaclast deposits are linked to an insufficiently high resolution of satellite images, as well as 'masking' effects of vegetation cover and cliff shadows. However, these limitations do not diminish the importance of the Google Earth Engine for finding these deposits. Consideration of some tourism-related information, including photos captured by tourists and bouldering catalogues, facilitates search for promising areas for subsequent virtual surveying of megaclast distribution. It is also established that the Google Earth Engine permits quantitative analysis of composition of coastal megaclast deposits in some areas, as well as to register decade-long dynamics or stability of these deposits, which is important to interpret their origin. The current opportunities for automatic detection of coastal megaclast deposits seem to be restricted.

**Keywords:** large clasts; remote sensing; bouldering tourism; Iberian Peninsula; Mediterranean; Indonesia; Central America

---

## 1. Introduction

Coastal sediment dynamics is always highly complex, and, thus, it provides a lot of issues for geoscience investigations. An interest in nomenclature development for large clasts that started near the end of the 20th century [1] and attention to coastal hazards facilitated by the Indian Ocean Tsunami of 2004 [2] have shaped a new international research direction, namely megaclast studies [3]. Megaclasts of ocean and sea coasts have been investigated most intensively. These studies have been conducted in so different places of the world as Baja California in Mexico [4,5], North Eleuthera of the Bahamas [6,7], and Rabat in Morocco [8]. Although the works have tended to focus on only some regions like the Mediterranean (e.g., [9–13]), the global evidence of large clast accumulations has become huge already, and it continues growing. This evidence requires extension and generalization for further conceptual treatment. Previous reviews of rocky shores [14] and megaclasts [3] confirm such research is promising. Undoubtedly, coastal megaclast deposits of Quaternary age are of special importance because the evidence of such deposits from the earlier geological periods remains scarce [3,15].

Coastal megaclast deposits attract significant attention of modern researchers for two aspects. The first aspect is the development of large clast nomenclature. Somewhat coherent and somewhat alternative proposals were made by Blair and McPherson [1], Bruno and Ruban [16], Blott and Pye [17], and Terry and Goff [18]. Additionally, Cox et al. [10] offered a novel approach for measurements of roundness of megaclasts, and, thus, the noted nomenclature development should emphasize not only on grain-size categories, but also on various morphological parameters. The second aspect of coastal megaclast deposits is linked to genetic investigations and, particularly, interpretations of the evidence of past storms and tsunamis (storm-versus-tsunami origin of megaclasts is a popular and hotly-debated topic in the modern sedimentology) [4–13,19–28]. Irrespective of which of these aspects and particular opinions to follow, finding new localities coastal megaclast deposits is of utmost importance. The knowledge from the 'classical' localities like those studied in the Bahamas [6,7] and western Ireland [29–35] need extension and refinement with the information from many other localities of the world. Although megaclast studies are urgent and relatively cheap, the circle of the involved researchers remains too restricted to expect documentation of even a triple of all megaclast localities. Moreover, some of the latter can be situated in remote places travelling to where faces serious difficulties in regard to researchers time, safety, and expenses.

The dilemma of the high demand for the really global knowledge of megaclasts contrasting the geographical restriction of their research can be addressed with application of the modern remote sensing techniques. Particularly, the Google Earth Engine [36,37] seems to be promising due to by definition a big size of megaclasts that makes these well visible on satellite images. The extraterrestrial investigations of megaclasts have proven efficacy of similar approaches [16]. The objective of the present paper is to demonstrate the application of the Google Earth Engine for finding coastal megaclast deposits on the basis of several representative examples. This experience permits also to highlight perspectives and restrictions of this approach. In the other words, the focus of this paper is methodological, and it concerns chiefly virtual identification of localities, not comprehensive sedimentological description of deposits represented there. In this paper, it is also undertaken to summarize briefly the available information on megaclasts in some major regions of the Earth in order to demonstrate how finding new localities can fill geographical gaps in the knowledge of megaclast distribution along coasts.

## 2. Methodology

Large clasts are detrital sedimentary particles larger than 256 mm in size, according to the standard Udden–Wenthworth classification, or larger than 100 mm, according to the alternative classification proposed by Bruno and Ruban [16]. Boulders are large clasts with the size ranging between 0.10 and 1 m, and megaclasts are detrital sedimentary particles larger than 1 m in size [16]. However, it should be noted that different researchers proposed the different lower limit for this category of particles (= the different upper limit of boulders) [1,17,18]. Depending on their size, megaclasts can be subdivided into blocks (1–10 m), megablocks (10–100 m), and superblocks (>100 m) [16]. Coastal boulder deposits are distributed along coasts of seas, oceans, and great lakes and include a significant amount of large clasts (true boulders and megaclasts); these deposits associate often with rocky shores and reflect influences of storms and tsunamis and the relevant clast transport (inland, above high-water mark, and even to the cliff-top position) [13,20,27]. Synonymous terms are boulderite coined by Dewey and Ryan [21], boulder beach [19,38–40], and boulder field [8,41]. Finally, coastal megaclast deposits are coastal boulder deposits dominated by clasts larger than 1 m in size or, at least, bearing recognizable accumulations of such clasts.

The Google Earth Engine is a software that offers satellite images of different scales for the planetary surface and allows their analysis; it also incorporates well-justified cartographical basis, GIS technologies, and some other information, including photos provided by the users. This instrument can be applied successfully for solution of various geoscience tasks [36,37]. For instance, it has been used efficiently in landslide susceptibility mapping [42,43]. The most elementary function of the Google Earth Engine is visual surveying of the Earth surface at an appropriate resolution. Taking into account that megaclasts are >1 m in size, the resolution of the available satellite images permits finding them in almost all regions and describing them preliminarily when image resolution is especially high. A significant advantage of this approach is a no-cost, quick, and geographically unrestricted search for megaclast localities, including those in difficult-to-access places. Anyway, the efficacy of the Google Earth Engine in the search for coastal megaclast deposits needs testing, and the latter was addressed in the present study. The Google Earth Pro 7.3.2.5776 version (free software) of the Google Earth Engine was employed for this purpose. Images of the maximum appropriate resolution are preferred. In some cases, resolution can be increased a bit else, but this leads to smooth contours of natural objects, i.e., to poor-visibility of megaclasts.

In order to test the application of the Google Earth Engine to coastal megaclast deposits, four localities of the latter were considered. These were found as a result of tentative visual coastline surveying with the Google Earth Engine in those areas where the occurrence of megaclasts seems to be possible, but has not been reported earlier. The localities were Hondarribia in northern Spain (the Biscay Bay coast), the Ponza Island of Italy (Tyrrhenian Sea), the Wetar Island in eastern Indonesia (Banda Sea), and the Humboldt o Coredo Bay and vicinities near the border between Colombia and Panama (eastern Pacific coast; Figure 1). These localities represent different geographical domains, namely Atlantic Europe, the Mediterranean, Southeast Asia, and Central America. For each locality, two plots are selected provisionally for more representative approach testing. When possible, these plots were selected so to demonstrate local differences of coastal megaclast deposits. Although this paper focused on finding localities, preliminary qualitative descriptions of the studied coastal megaclast deposits were provided, as this study explored the very possibility of their 'visibility' with the Google Earth Engine. The descriptions avoid details relevant to remote sensing techniques, but emphasize on the general character of the deposits, which is the primary interest of sedimentologists. In other words, the descriptions were addressed to sedimentologists, not specialists in remote sensing. Importantly, these localities were selected to demonstrate the utility of satellite images with a different resolution.

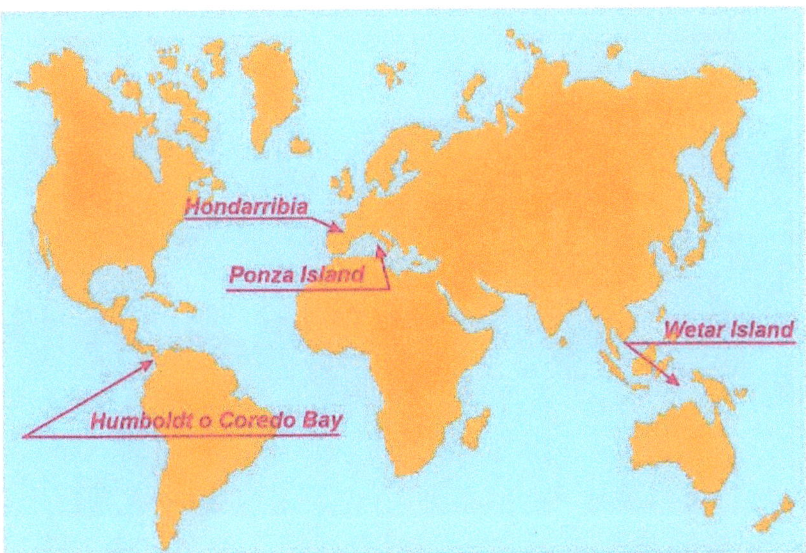


**Figure 1.** New localities of coastal megaclast deposits considered in this paper.

## 3. Results

Four small case studies were undertaken in order to demonstrate the efficacy of the virtual examination of coastal megaclast deposits with the Google Earth Engine. Two examples deal with satellite images of exceptionally high resolution, and two other examples deal with satellite images of appropriate but somewhat limited resolution. In each case, it is also attempted to stress the relative importance of the new evidence to the regional knowledge of coastal megaclast deposits. For this purpose, the basic published information was summarized.

### 3.1. Example 1: Hondarribia (Spain; Biscay Bay)

The locality is situated in the vicinities of the town of Hondarribia in northern Spain (Figure 1), where the rocky shore of the Biscay Bay hosts a lot of megaclasts that form significant accumulations. Their origin is linked to the destruction of the Paleogene marine rocks exposed in cliffs and directly along the coastline [44]. Apparently, the main factors of megaclast accumulation are gravity processes and wave abrasion.

On the eastern plot, the coastal megaclast deposits were represented by the dense accumulation of angular blocks with the size of 1–5 m; the maximum size reached 7–8 m (Figure 2). Importantly, the deposits were spatially heterogeneous—the mean size of blocks increased locally (for instance, near the shoreline and close to parent rock exposures). Although the deposits covered almost the entire area, some 'islands' of parent rock exposures and semi-detached megaclasts were also visible. On the western plot, angular megaclasts were larger in size: these include blocks and megablocks; the size of the largest particle exceeded 25 m (Figure 3). Megablocks were more numerous in the right half of the plot. Parent rock exposures and semi-detached megaclasts occurred in the left half. On both plots, clasts >1 m in size were distinguished unequivocally, and the evidence was enough to confirm the presence of true coastal megaclast deposits. The resolution of the satellite view permitted us to distinguish blocks from megablocks; the presence of boulders was evident, although these could not be registered individually. Dense package of the deposits was well visible. Size and shape of individual blocks can be registered when their size exceeds 1–2 m. The spatial position and orientation of stones did not leave an impression of significant re-working, i.e., primary, undisturbed coastal megaclast deposits could be hypothesized.

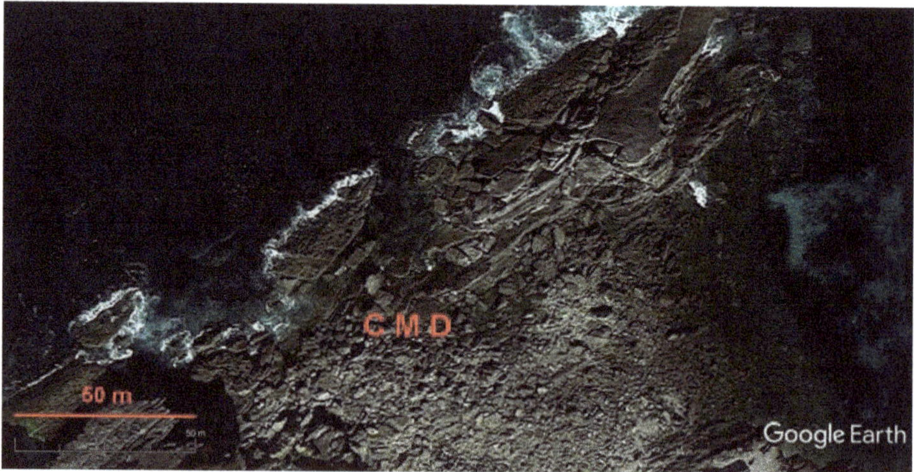


**Figure 2.** The eastern plot of the Hondarribia locality (the view provided by the Google Earth Engine). CMD labels coastal megaclast deposits (the same is on the following figures).

**Figure 3.** The western plot of the Hondarribia locality (the view provided by the Google Earth Engine).

The information from this locality fills an important gap in the knowledge of the geographical distribution of coastal megaclast deposits along the Atlantic coast of Europe. These deposits were reported previously from the British Isles [21,27–35,45–51], northern France (first of all, Brittany) [52–56], and the western and southwestern coasts of the Iberian Peninsula [57–61]. The Hondarribia locality represents the other domain, i.e., the Biscay Bay, the coasts of which are prone to severe storms creating and transporting megaclasts, similarly to the above-mentioned regions [62]. This may be the first megaclast locality reported from the Biscay Bay coast.

### 3.2. Example 2: Ponza Island (Italy; Tyrrhenian Sea)

The locality corresponds to the Ponza Island (Pontine Archipelago) that is situated in the Tyrrhenian Sea, near the western coast of central Italy (Figure 1). The Pliocene volcanic and volcaniclastic rocks are exposed in tall cliffs [63–65], the collapse of which leads to the formation and accumulation of numerous megaclasts. As rocky shores with cliffs dominate the island, the distribution of coastal megaclast deposits is significant there. Gravity processes and abrasion contribute to local megaclast formation.

On the southern plot, the coastal megaclast deposits were represented by the accumulation of angular blocks with the common size of 1–5 m (Figure 4). Blocks of a bigger size and rare small megablocks (up to 12–13 m in size) occurred along the very shoreline, as well as in the northern part of the plot. Density of the deposit differed: it was significant near the shoreline, but megaclasts occurred individually or in small groups in the other places where they were either mixed with boulders or lay directly on the rock surface. On the northwestern plot, a small group of fully- and partially-detached megaclasts was observed (Figure 5). These include several megablocks with the size up to 40 m and numerous blocks in between. These megaclasts were angular, and some smoothed surfaces result from wave and wind 'polishing'. On both plots, clasts >1 m in size were distinguished easily, and the delineation of true coastal megaclast deposits was easy. The resolution of satellite images permitted us to distinguish blocks from megablocks, the presence of boulders was evident, although these could not be registered individually. Different package density was visible. Size and shape of individual particles can be registered when their size exceeded 1–2 m. However, a shadow from the tall cliff on the northwestern plot 'masked' some megaclasts and did not permit us to characterize them on an individual basis (Figure 5). Some redeposition of megaclasts could not be excluded for the southern plot, but the view of the deposits on the northwestern plot leaves impression of a 'fresh' and then undisturbed rockfall.

**Figure 4.** The southern plot of the Ponza Island locality (the view provided by the Google Earth Engine).

**Figure 5.** The northwestern plot of the Ponza Island locality (the view provided by the Google Earth Engine).

The Mediterranean is a 'classical' region for studies of coastal megaclasts and boulder deposits [9–13]. Particularly, megaclast accumulations have been reported from the Western Mediterranean continental and island coasts [66–68]. However, Ponza Island is of special importance because of two reasons. First, it represents lenticular, almost round-island distribution of the coastal megaclast deposits. Second, the latter originates from volcanic and volcaniclastic rocks. Therefore, finding this locality with the Google Earth Engine extends the regional knowledge of coastal megaclast deposits.

### 3.3. Example 3: Wetar Island (Indonesia; Banda Sea)

The locality corresponds to the Banda Sea coast of the Wetar Island in eastern Indonesia (Figure 1). There, Cenozoic volcanic, volcaniclastic, and reefal carbonate deposits crop out [69]. Presumably, their destruction by wave abrasion and weathering leads to megaclast creation; it cannot be excluded that some megaclasts were formed as a result of slope collapse and subsequent downslope transport. This means that the origin of this new locality can be highly complex, but the role of abrasion seems to be leading.

On the eastern plot, blocks with the maximum size of up to 10 m (the prevailing size was 3–5 m) occurred sporadically (Figure 6). These lay either individually or in small groups. The biggest stones were angular. The deposit package density was very low with regard to the distance between megaclasts (even when these occur in groups). These coastal megaclast deposits were restricted to the very shoreline. On the northeastern plot, the number of megaclasts was bigger (Figure 7). These were chiefly blocks of different size (commonly < 5 m), but there were also megablocks, the biggest of which reached 20 m in size. The particle angularity was well visible. The package density was moderate (even high in some places), but the deposits also occurred like a narrow ribbon along the shoreline. On the both plots, clasts >3 m in size were distinguished easily; the satellite image evidence is enough to confirm the presence of true coastal megaclast deposits. The resolution of the images permitted us to distinguish blocks from megablocks; boulders were not seen at all (it was unclear whether these exist). A different package density was visible without any difficulty. Size and shape of individual particles can be registered when their size exceeds 5 m. Better to say, these can be examined chiefly for large blocks and all megablocks. Dense vegetation cover of the island and, particularly, along the shoreline 'masks' the coastal megaclast deposits, but this does not preclude for megaclast accumulation tracing over the studied plots. The position and the orientation of megaclasts do not suggest against their resedimentation, and, if so, it cannot be excluded that high-energy events contributed to motion and destruction of these stones.

**Figure 6.** The eastern plot of the Wetar Island locality (the view provided by the Google Earth Engine).

Coastal megaclasts have been reported earlier from many localities of Southeast Asia, including those of China, Taiwan, Thailand, Malaysia, Indonesia, and the Philippines [70–94]. Strong influences of tsunamis [95,96] and super-typhoons [97] contribute to the formation of such deposits. However, all this evidence has been obtained outside eastern Indonesia and chiefly on coasts open to either the Indian or Pacific Oceans. The Wetar Island locality sheds light on the coastal megaclast deposits formed in a distinct geographical setting, namely on the coast of a relatively small, intra-island sea, which makes addition to the regional knowledge of such deposits. Therefore, the Google Earth Engine permits extension of the available knowledge of megaclast occurrence in Southeast Asia.

**Figure 7.** The northeastern plot of the Wetar Island locality (the view provided by the Google Earth Engine).

*3.4. Example 4: Humboldt o Coredo Bay (Colombia/Panama; Eastern Pacific)*

The locality corresponds to the Humdoldt o Coredo Bay and its vicinities. It is situated on the Pacific coast of Central America, exactly at the border between Colombia and Panama, north of Jurado (Figure 1) where the Cretaceous-Eocene basement and volcanic arc complex dominate the local geological setting [98]. Wave destruction and weathering of hard parent rocks contributes to formation of multiple megaclasts. Apparently, abrasion is chiefly responsible for the origin of these deposits.

On the southern plot, small and medium blocks (typical size is <5 m) formed lenticular deposits around a small bay (Figure 8). The biggest particles were angular, but the shape of the majority of particles could not be recognized at the available resolution. The width of the ribbon of deposits differed, as well as their package density. A few huge blocks (but these did not reach the size of megablocks) lay individually in the central part of the plot. Generally, it is possible to record the deposits heterogeneity. On the northern plot, the coastal megaclast deposits were very similar to those described on the southern plot, with two exceptions: the mean size of blocks was bigger, and there were a few megablocks (up to 20 m in size or even more; Figure 9). On both plots, clasts >3 m in size were distinguished easily, and the presence of true coastal megaclast deposits was evident. The resolution of satellite images permitted us to distinguish blocks from megablocks; boulders were not seen at all, although these, presumably, exist. A different package density was visible without any difficulty. The size and shape of individual particles can be registered when their size exceeds 5 m (large blocks and megablocks). The difficulties were linked, first, to the 'masking' effect of the vegetation cover and, second, uncertainty with some features on the northern plot that could be either megaclasts or parent rock exposures (or semi-detached megaclasts). Presumably, the both difficulties will remain even in the case of a much higher resolution of satellite images. The spatial position and orientation of stones permitted us to hypothesize the absence of significant reworking, i.e., these seemed to be primary, undisturbed coastal megaclast deposits.

Previous megaclast studies on the Pacific coast of the Americas are scarce, and, particularly, focused on two geographical domains, namely the Baja California Peninsula in Mexico [4,5] and northern and central Chile [41,99,100]. If so, the information from the Humboldt o Coredo Bay locality fills significant gap in the regional knowledge of coastal megaclast deposits characterizing those of Central America. Finding this locality with the Google Earth Engine implies the existence of coastal megaclast deposits on the Pacific coast of Central America (probably, for the first time).

**Figure 8.** The southern plot of the Humboldt o Coredo Bay locality (the view provided by the Google Earth Engine).

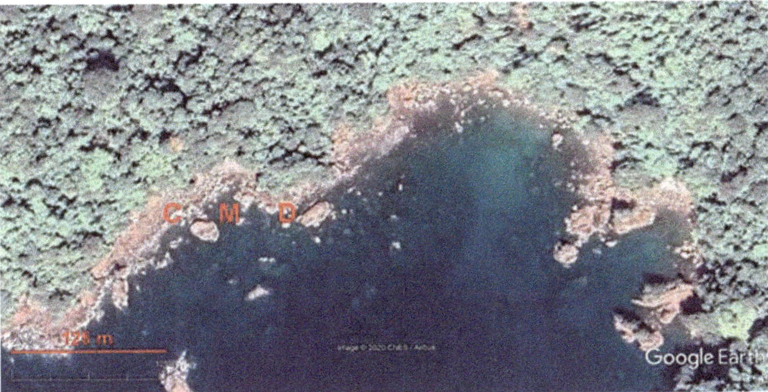

**Figure 9.** The northern plot of the Humboldt o Coredo Bay locality (the view provided by the Google Earth Engine).

## 4. Discussion

### 4.1. Synthesis: Virtual Approach Opportinities and Limitations

The four examples given above illustrate the application of the Google Earth Engine to the identification and even preliminary description of coastal megaclast deposits. The approach seems to be promising. In all four cases, the deposits were identified and delineated without any difficulty and irrespective to the different resolution of satellite images. All study areas were new to megaclast research, and these were found on the basis of the only virtual exploration of the coastline, without consideration of any preliminary evidence. This approach can be used for visual surveying of the coasts and identification of localities suitable for subsequent field or virtual investigations (or both). Moreover, some basic characteristics of coastal megaclast deposits can be examined with the Google Earth Engine taken alone, although such a study cannot 'replace' in-depth field investigations to be undertaken further. These characteristics include the dominant particle size, package density, and spatial homogeneity versus heterogeneity of deposits. Individual, group, or lenticular occurrence of megaclasts can be also registered. As for the particle size, in all cases, it was not only possible to make a distinction of blocks from megablocks, but also to note the presence of smaller and larger blocks, i.e., to deal with subcategories of the main grain-size divisions. Some notions on the possible origin of the deposits are also possible. Importantly, the use of the Google Earth Engine permitted us a very quick survey of coasts. Apparently, one researcher can check up to a thousand kilometers of coastal zone for the presence of megaclast accumulations per 5–7 working hours. No field investigation can allow such a speed of work and so massive information gathering. Moreover, many areas will remain 'untouched' by megaclast research due to the relatively circle of the involved researchers and logistic restrictions. If so, it is better to know the position of the localities and to have the only preliminary characteristics of the deposits than to concentrate on only well-known and well-accessible localities studied for years.

As implied by the examples, virtual studies of coastal megaclast deposits with the Google Earth Engine face limitations of three kinds. First, the resolution of satellite images did not permit us to deal with boulders and other finer components of the deposits, and the presence of these components could not be proven in some cases. Second, the resolution of the satellite images available for the Hondarribia and Ponza Island localities allowed the size and shape documentation of all megaclasts, whereas the resolution of the satellite images of the Wetar Island and Humboldt o Coredo Bay localities permitted doing this for only huge blocks and megablocks, i.e., for the only part of the deposit. In the two latter cases, there were also difficulties with distinction between true megaclasts, semi-detached megaclasts, and exposures of parent rocks. This means that detailed investigations of coastal megaclast deposits and their quantitative descriptions with the Google Earth Engine are possible in only some cases. Third, there are local limitations of the approach like tall cliff shadows (the Ponza Island) or dense vegetation cover (Wetar Island and Humboldt o Coredo Bay). Anyway, all these limitations do not preclude from the documentation of the spatial occurrence of megaclast deposits along coasts of seas and oceans. When necessary, field studies can be undertaken later, and these studies can be coordinated effectively with the information obtained with the Google Earth Engine. In other words, the limitations are less important than the opportunities. Moreover, it is expected that the resolution of satellite images employed by the noted software will continue to increase in the nearest future.

*4.2. Tourism Information as a Source for Geographical Justification of Virtual Surveys*

The total world coastline is too lengthy taking into account continental and all island coasts, and even the Google Earth Engine does not make efficient its visual surveying for the purposes of the identification and description of megaclast deposits. The latter occurs locally, and it is sensible to focus on those areas, which seem to be promising for finding megaclasts. Some previous descriptions ('occasional' descriptions of megaclast deposits can be found in the literature) and geological, geomorphological, and geographical knowledge (for instance, the selection of coasts with tall cliffs and prone to severe storms, super-typhoons, and tsunamis) is helpful in many cases for the geographical justification of virtual surveys. However, some other, non-scientific evidence can be also considered.

The first opportunity is linked to photos provided by the users (first of all, tourists) of the Google Earth Engine, the geographical attachment of which is displayed directly on satellite images. Megaclasts and their accumulations are notable features, and these can be recognized easily on photos. Moreover, the latter can help in satellite image interpretation, i.e., for the correct description of the particle shape and package density, as well as for the distinction between true megaclasts, semi-detached megaclasts, and parent rock exposures; the content of boulders and finer components of deposits can be also registered. In the present study, the photos provided by the Google Earth Engine were helpful in the cases of the Hondarribia and Ponza Island localities. However, the number of the proper photos was limited for some remote, infrequently visited places, or these were not available at all.

The second opportunity that can be recommended for the geographical inventory of coastal boulder deposits is also relevant to tourism-linked information available online. Among various sport and tourism activities, bouldering has gained importance [101–105]. A lot of information on individual 'boulders' (these may be either isolated rocks, but also true megaclasts) has been accumulated. This information is easily accessible with the available on-line catalogues [106–108], and it permits finding coastal megaclasts and their groups, as well as their preliminary visual examination with the available images. For instance, there are typical coastal boulder deposits on the Ao Sane Beach in Phuket, Thailand [106]. These deposits consist of mixed angular blocks and boulders, and there are also angular blocks lying separately on a sandy substrate. In regard to the position of the later, resedimentation cannot be excluded. Such information can be of the utmost importance for finding promising areas to focus virtual surveys of coastal megaclast deposits.

### 4.3. Do Satellite Images Provide Data for the Quantitative Analysis of Coastal Megaclast Deposits?

Finding new localities of coastal megaclast deposits with the Google Earth Engine poses a question of more detailed analysis providing sedimentological information. The examples given above imply that some descriptions of such deposits are possible, but these are preliminary and too qualitative. Actually, in some cases, quantitative interpretations seem to be possible. For instance, grain-size distribution of megaclasts can be analyzed with statistical tools. There are four conditions that allow such an investigation. First, the resolution of satellite images of the study area should allow unequivocal recognition of particles with the size of 1–2 m (minimal size of megaclasts). Second, there should be not any 'masking' effects of shadows, vegetation, etc. Third, the majority of megaclasts should be oriented subhorizontally because vertical orientation does not permit the measurement of the maximum size. Fourth, megaclasts should form a relatively thin layer' alternatively, the lowermost particles covered by the other blocks cannot be measured.

From the four examples considered in this study, the only Hondarribia locality satisfied all four conditions outlined above. To demonstrate the efficacy of satellite image analysis, the western plot of the locality (Figure 3) was considered. The Google Earth Engine provides the option of the measurement of objects directly on a satellite image. The maximum size of 100 megaclasts was measured with this tool on the chosen plot. Apparently, the measured particles constituted about 70% of all megaclasts visible on this plot. The size of megaclasts varied between 1.0 and 26.2 m. The mean size was 7.8 m, and the median size was 6.3 m. These values corresponded to coarse blocks, according to the classification of Bruno and Ruban [16]. The grain-size distribution of megaclasts is shown on a histogram (Figure 10). The majority of particles had the size of 2.5–5 m, i.e., these were medium blocks. However, coarse blocks (5–10 m) and fine megablocks (10–25 m) were the most common. Generally, these coastal boulder deposits included 37% of coarse blocks, 26% of medium blocks, 24% of fine megablocks, 12% of fine blocks, and 1% of medium megablocks, i.e., it was dominated by the particles with the size of 2.5–25 m, which indicates on a restricted sorting. Undoubtedly, this deposit also bore some finer components like boulders, gravels, and even sand, but these were invisible on the satellite image. Anyway, the dominance of the noted megaclasts in these deposits was indisputable.

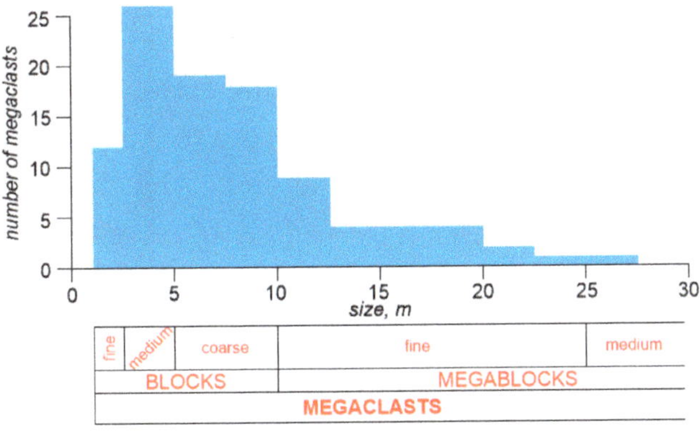


**Figure 10.** Grain-size distribution of megaclasts on the western plot of the Hondarribia locality.

*4.4. Satellite Evidence of a Decade-Long Stability of Coastal Megaclast Deposits*

The modern studies of megaclast focus much (even over-emphasize) on their origin and, particularly, their relevance to extreme events such as storms and tsunamis [4–13,19–27]. Evidently, these studies deal with the dynamics of the coastline and megaclasts themselves. The Google Earth Engine gives exceptional opportunity to contribute to such investigations, as a time series of satellite images is available. This means that direct comparison of the views of some areas at different time slices is possible. Three limitations are as follows. First, 'old' satellite images are not available for all areas and the date of these images differs for different areas. Second, the resolution of 'old' images can be lower than that of the current images. Third, due to a different time of image making, shadows, vegetation, and, importantly, water level may look differently. However, the third problem can be overcome in many cases.

In order to understand the importance of the analysis of the image time series for the understanding of the dynamics of coastal boulder deposits, the new localities identified in the present work were taken as examples. 'Old' satellite images (captured 10–20 years ago) of appropriate resolution are available for three of them (Hondarribia, Ponza Island, and Wetar Island—one plot was selected for analysis in each case). However, in the only case of the Ponza Island, the maximum appropriate resolution of the 'old' image was the same as that of the 'new' image. In two other cases (Hondarribia and Wetar Island), the resolution of the 'old' image was lower, but it allowed recognition of the principal features, i.e., megaclasts, via comparison to the higher-resolution 'new' image. Megaclasts from the western plot of the Hondarribia locality occupied the same position in 2001 (Figure 11) as in 2018 when the currently available image was captured (Figure 3). Apparently, the number of blocks and megablocks did not change, although some of them had worse visibility on the 'old' image because of the higher sea level (apparently, the time of tide). No changes were also found between 2007 (Figure 12) and 2017 (Figure 5) on the northwestern plot of the Ponza Island. Moreover, the 'old' image was of better quality in this case. The shadow from a tall cliff did not 'mask' a part of the deposits, and the better seawater transparency permitted us to document that up to a quarter of the deposits stretch to the submarine environment. Finally, nothing related to megaclasts changed on the eastern plot of the Wetar Island between 2009 (Figure 13) and 2019 (Figure 6).

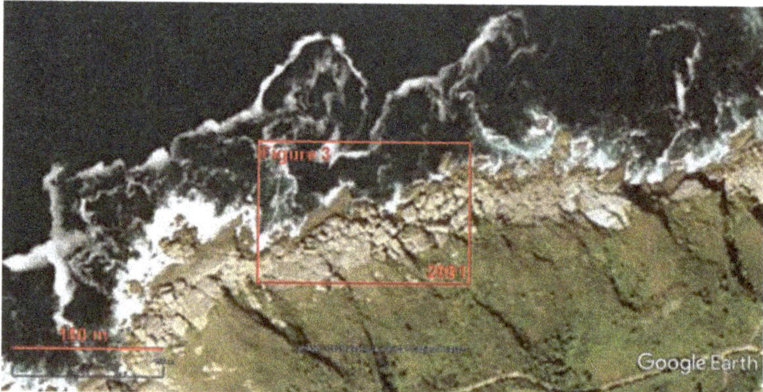


**Figure 11.** Satellite view of the western plot of the Hondarribia locality in 2001 (the view provided by the Google Earth Engine).

**Figure 12.** Satellite view of the northwestern plot of the Ponza Island locality in 2009 (the view provided by the Google Earth Engine).

**Figure 13.** Satellite view of the eastern plot of the Wetar Island locality in 2007 (the view provided by the Google Earth Engine).

The evidence presented above implies a decade-long stability of the coastal megaclast deposits in the new localities. Megaclasts were not destroyed, replaced, or over-turned despite that the regions are susceptible to severe weather conditions [62], and tsunamis cannot be excluded there [109,110]. Even if the periodicity of the high-energy events that can be responsible for changes of coastal megaclast deposits seems to be longer than a decade, the registered stability is notable. This is especially the case of the Hondarribia locality located on a high-energy coast of the Biscay Bay, for which a two-decade long comparison (Figures 3 and 11) is possible. Apparently, severe storms of the 2000s and the 2010s [62] did not affect megaclasts of this locality.

It is sensible to add that the Google Earth Engine gives opportunity to get images with a high frequency (Sentinel 2 or SPOT 7 images). If so, some localities can be specially monitored in the case of any future severe storms or tsunamis to document the dynamics of coastal megaclast deposits.

*4.5. Automatic Detection of Coastal Megaclast Deposits: the Current State of the Problem*

The development of the Google Earth Engine has permitted to pose the question of the invention of advanced digital tools for automatic detection of particular features on satellite images. For instance, such tools are helpful in urban studies [111] and archaeology [112]. In geomorphological studies, two successfully tested approaches are notable. The first has been proposed by Luijendijk et al. [113]

for the global-scale mapping of sandy beaches and establishing their dynamics. The second approach has been employed by Vos et al. [114] to document shoreline dynamics (some examples also deal with sandy beaches). The validity and the importance of the both above-mentioned studies are indisputable, and the availability of any similar approach would help in the global mapping of coastal megaclast deposits. It is not the purpose of the present work to develop such an advanced interpretative technique, but some limitations of the latter can be discussed in the light of the present findings.

First of all, the main differences of coastal megaclast deposits from sandy beaches should be noted. These deposits are more localized and often narrow; these rarely form lengthy 'belts' like sandy beaches. The latter are often distinguished on satellite images by a white or yellow color, whereas coastal megaclast deposits may look very differently in each case. Importantly, these deposits are often not continuous particle packages, but individual stones or stone groups dispersed along the coastline (this is especially well visible at the Weater Island locality—Figures 6 and 7). To make a clear distinction between true megaclasts and semi-detached megaclasts or rock exposures is not always easy, especially when the surface of a given area is not flat. When sandy beaches are detected, the objects of study are beaches themselves, and the researchers do not to pay attention to sand particles that cannot be determined from the space. In contract, studies of coastal megaclast deposits require attention to giant particles. That is why such studies depend stronger on image resolution. The experience with the Ponza Island (Figures 5 and 12) implies water transparency influences on the detection of megaclasts that are fully or partly drowned, whereas the evidence from the Hondarribia locality (Figures 3 and 11) indicates on the importance of the sea level, i.e., tides. As a result, visibility of coastal megaclast deposits decreases under certain conditions (Figure 14).

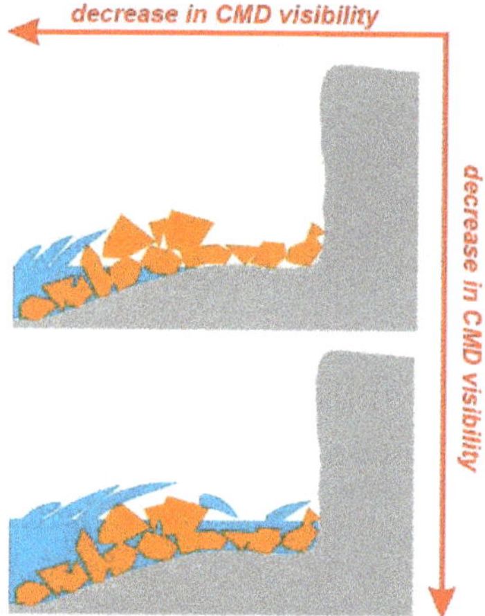


**Figure 14.** Dependence of coastal megaclast deposit visibility on satellite images on the sea-level position.

The above-said does not suggest against the necessity of techniques for automated detection of coastal megaclast deposits, but reveals serious barriers for the invention of such techniques. Moreover, it should be stressed that the studies of Luijendijk et al. [113] are aimed at making global- or regional-scale conclusions. In contrast, what do the megaclast researchers actually need are particular localities, especially providing representative examples of coastal megaclast deposits from different parts of the world. In regard to this, simple visual surveys of shorelines with the Google Earth Engine provide reasonable information.

## 5. Conclusions

The use of the Google Earth Engine has permitted to us find four new localities of coastal megaclast deposits in Atlantic Europe, the Mediterranean, Southeast Asia, and Central America. The dominance of blocks and subordinate number of megablocks, as well as their chiefly angular shape are visible on satellite images. More detailed, quantitative investigations are possible on the basis of the both field and virtual studies of these localities. Despite some limitations, the Google Earth Engine seems to be an almost ideal instrument for quick tracing the geographical distribution of megaclast deposits along the coasts of seas and oceans, which is important for the coordination of further research. The growth of tourism and voluminous online representation of the relevant photos facilitate finding areas promising for identification of megaclast localities. The novelty of this study is triple, i.e., the potential of virtual search for coastal megaclast deposits was proven, four new localities found, and the knowledge of megaclast spatial occurrence in several large regions was summarized (it was also explained how geographical gaps in this knowledge could be filled with virtual finding of new localities). More generally, this work did not argue for 'replacing' field studies by virtual surveys. In contrast, it demonstrated how to find efficiently new localities for further field studies. In the other words, finding localities as a research task is separated from in-depth sedimentological investigations.

This paper dealt with four, almost randomly selected localities. Further virtual surveys of the coastal zones will permit to extend the relevant knowledge, as well as to improve the methods of the satellite image analysis. One of the central problems is the precise megaclast size and shape description, as satellite images show the only 2D projection of these features, which can be oriented differently. It is important problem to think how some tools proposed earlier [115–117] can be coupled with the use of the Google Earth Engine in coastal megaclast studies. It cannot be excluded that satellite images can be used for automatic detection of megaclast-promising areas in the future.

**Author Contributions:** D.A.R. is the only author of this work, and he is fully responsible for its content. The Author has read and agree to the published version of the manuscript.

**Funding:** This research received no external funding.

**Acknowledgments:** The author gratefully thanks M.E. Johnson (USA) for his kind invitation to contribute to this special issue and various support, as well as all reviewers for their helpful suggestions.

**Conflicts of Interest:** The author declares no conflict of interest.

## References

1. Blair, T.C.; McPherson, J.G. Grain-size and textural classification of coarse sedimentary particles. *J. Sediment. Res.* **1999**, *69*, 6–19. [CrossRef]
2. Ruban, D.A. Research in tsunami-related sedimentology during 2001–2010: Can a single natural disaster re-shape the science? *GeoActa* **2011**, *10*, 79–85.
3. Ruban, D.A.; Ponedelnik, A.A.; Yashalova, N.N. Megaclasts: Term Use and Relevant Biases. *Geosciences* **2019**, *9*, 14. [CrossRef]
4. Johnson, M.E.; Ledesma-Vazquez, J.; Guardado-France, R. Coastal Geomorphology of a Holocene Hurricane Deposits on a Pleistocene Marine Terrace from Isla Carmen (Baja California Sur, Mexico). *J. Mar. Sci. Eng.* **2018**, *6*, 108. [CrossRef]
5. Johnson, M.E.; Guardado-France, R.; Johnson, E.M.; Ledesma-Vazquez, J. Geomorphology of a Holocene Hurricane Deposit Eroded from Rhyolite Sea Cliffs on Ensenada Almeja (Baja California Sur, Mexico). *J. Mar. Sci. Eng.* **2019**, *7*, 193. [CrossRef]
6. Hearty, P.J.; Tormey, B.R. Sea-level change and superstorms; geologic evidence from the last interglacial (MIS 5e) in the Bahamas and Bermuda offers ominous prospects for a warming Earth. *Mar. Geol.* **2017**, *390*, 347–365. [CrossRef]
7. Mylroie, J.E. Superstorms: Comments on Bahamian Fenestrae and Boulder Evidence from the Last Interglacial. *J. Coast. Res.* **2018**, *34*, 1471–1483. [CrossRef]

8.    Medina, F.; Mhammdi, N.; Chiguer, A.; Akil, M.; Jaaidi, E.B. The Rabat and Larache boulder fields; new examples of high-energy deposits related to storms and tsunami waves in north-western Morocco. *Nat. Hazards* **2011**, *59*, 725. [CrossRef]

9.    Biolchi, S.; Denamiel, C.; Devoto, S.; Korbar, T.; Macovaz, V.; Scicchitano, G.; Vilibic, I.; Furlani, S. Impact of the October 2018 storm Vaia on coastal boulders in the northern Adriatic Sea. *Water* **2019**, *11*, 2229. [CrossRef]

10.   Causon Deguara, J.; Gauci, R. Evidence of extreme wave events from boulder deposits on the south-east coast of Malta (Central Mediterranean). *Nat. Hazards* **2017**, *86*, 543–568. [CrossRef]

11.   Scheffers, A.; Scheffers, S. Tsunami deposits on the coastline of west Crete (Greece). *Earth Planet. Sci. Lett.* **2007**, *259*, 613–624. [CrossRef]

12.   Scicchitano, G.; Monaco, C.; Tortorici, L. Large boulder deposits by tsunami waves along the Ionian coast of south-eastern Sicily (Italy). *Mar. Geol.* **2007**, *238*, 75–91. [CrossRef]

13.   Shah-Hosseini, M.; Saleem, A.; Mahmoud, A.-M.A.; Morhange, C. Coastal boulder deposits attesting to large wave impacts on the Mediterranean coast of Egypt. *Nat. Hazards* **2016**, *83*, 849–865. [CrossRef]

14.   Johnson, M.E. Uniformitarianism as a guide to rocky-shore ecosystems in the geological record. *Can. J. Earth Sci.* **2006**, *43*, 1119–1147. [CrossRef]

15.   Ruban, D.A. Coastal Boulder Deposits of the Neogene World: A Synopsis. *J. Mar. Sci. Eng.* **2019**, *7*, 446. [CrossRef]

16.   Bruno, D.E.; Ruban, D.A. Something more than boulders: A geological comment on the nomenclature of megaclasts on extraterrestrial bodies. *Planet. Space Sci.* **2017**, *135*, 37–42. [CrossRef]

17.   Blott, S.J.; Pye, K. Particle size scales and classification of sediment types based on particle size distributions: Review and recommended procedures. *Sedimentology* **2012**, *59*, 2071–2096. [CrossRef]

18.   Terry, J.P.; Goff, J. Megaclasts: Proposed revised nomenclature at the coarse end of the Udden-Wentworth grain-size scale for sedimentary particles. *J. Sediment. Res.* **2014**, *84*, 192–197. [CrossRef]

19.   Chen, B.; Chen, Z.; Stephenson, W.; Finlayson, B. Morphodynamics of a boulder beach, Putuo Island, SE China coast: The role of storms and typhoon. *Mar. Geol.* **2011**, *283*, 106–115. [CrossRef]

20.   Cox, R.; O'Boyle, L.; Cytrynbaum, J. Imbricated Coastal Boulder Deposits are Formed by Storm Waves, and Can Preserve a Long-Term Storminess Record. *Sci. Rep.* **2019**, *9*, 10784. [CrossRef]

21.   Dewey, J.F.; Ryan, P.D. Storm, rogue wave, or tsunami origin for megaclast deposits in western Ireland and North Island, New Zealand? *Proc. Natl. Acad. Sci. USA* **2017**, *114*, E10639–E10647. [CrossRef]

22.   Hoffmann, G.; Grützner, C.; Schneider, B.; Preusser, F.; Reicherter, K. Large Holocene tsunamis in the northern Arabian Sea. *Mar. Geol.* **2020**, *419*, 106068. [CrossRef]

23.   Lorang, M.S. A wave-competence approach to distinguish between boulder and megaclast deposits due to storm waves versus tsunamis. *Mar. Geol.* **2011**, *283*, 90–97. [CrossRef]

24.   Noormets, R.; Crook, K.A.W.; Felton, E.A. Sedimentology of rocky shorelines: 3. Hydrodynamics of megaclast emplacement and transport on a shore platform, Oahu, Hawaii. *Sediment. Geol.* **2004**, *172*, 41–65. [CrossRef]

25.   Nott, J. Waves, coastal boulder deposits and the importance of the pre-transport setting. *Earth Planet. Sci. Lett.* **2003**, *210*, 269–276. [CrossRef]

26.   Scheffers, A.M.; Kinis, S. Stable imbrication and delicate/unstable settings in coastal boulder deposits: Indicators for tsunami dislocation? *Quat. Int.* **2014**, *332*, 73–84. [CrossRef]

27.   Williams, D.M.; Hall, A.M. Cliff-top megaclast deposits of Ireland, a record of extreme waves in the North Atlantic - Storms or tsunamis? *Mar. Geol.* **2004**, *206*, 101–117. [CrossRef]

28.   Cox, R.; Lopes, W.A.; Jahn, K.L. Quantitative roundness analysis of coastal boulder deposits. *Mar. Geol.* **2018**, *396*, 114–141. [CrossRef]

29.   Cox, R. Very large boulders were moved by storm waves on the west coast of Ireland in winter 2013–2014. *Mar. Geol.* **2019**, *412*, 217–219. [CrossRef]

30.   Cox, R.; Jahn, K.L.; Watkins, O.G.; Cox, P. Extraordinary boulder transport by storm waves (west of Ireland, winter 2013–2014), and criteria for analysing coastal boulder deposits. *Earth-Sci. Rev.* **2018**, *177*, 623–636. [CrossRef]

31.   Cronin, A.; Devoy, R.; Bartlett, D.; Nuyts, S.; O'Dwyer, B. Investigation of an elevated sands unit at Tralispean Bay, South-West Ireland—potential high-energy marine event. *Ir. Geogr.* **2018**, *51*, 229–260.

32.   Erdmann, W.; Kelletat, D.; Kuckuck, M. Boulder ridges and washover features in Galway Bay, Western Ireland. *J. Coast. Res.* **2017**, *33*, 997–1021. [CrossRef]

33. Erdmann, W.; Scheffers, A.M.; Kelletat, D.H. Holocene Coastal Sedimentation in a Rocky Environment: Geomorphological Evidence from the Aran Islands and Galway Bay (Western Ireland). *J. Coast. Res.* **2018**, *34*, 772–792. [CrossRef]
34. Scheffers, A.; Kelletat, D.; Haslett, S.; Scheffers, S.; Browne, T. Coastal boulder deposits in Galway Bay and the Aran Islands, western Ireland. *Z. Geomorphol.* **2010**, *54*, 247–279. [CrossRef]
35. Scheffers, A.; Scheffers, S.; Kelletat, D.; Browne, T. Wave-emplaced coarse debris and megaclasts in Ireland and Scotland: Boulder transport in a high-energy littoral environment. *J. Geol.* **2009**, *117*, 553–573. [CrossRef]
36. Gorelick, N.; Hancher, M.; Dixon, M.; Ilyushchenko, S.; Thau, D.; Moore, R. Google Earth Engine: Planetary-scale geospatial analysis for everyone. *Remote Sens. Environ.* **2017**, *202*, 18–27. [CrossRef]
37. Mutanga, O.; Kumar, L. Google Earth Engine applications. *Remote Sens.* **2019**, *11*, 591. [CrossRef]
38. Gómez-Pazo, A.; Pérez-Alberti, A.; Trenhaile, A. Recording inter-annual changes on a boulder beach in Galicia, NW Spain using an unmanned aerial vehicle. *Earth Surf. Process. Landf.* **2019**, *44*, 1004–1014. [CrossRef]
39. Green, A.; Cooper, A.; Salzmann, L. Longshore size grading on a boulder beach. *J. Sediment. Res.* **2016**, *86*, 1123–1128. [CrossRef]
40. Lorang, M.S. Predicting threshold entrainment mass for a boulder beach. *J. Coast. Res.* **2000**, *16*, 432–445.
41. Abad, M.; Izquierdo, T.; Caceres, M.; Bernardez, E.; Rodriguez-Vidal, J. Coastal boulder deposit as evidence of an ocean-wide prehistoric tsunami originated on the Atacama Desert coast (northern Chile). *Sedimentology* **2019**. [CrossRef]
42. Broeckx, J.; Vanmaercke, M.; Duchateau, R.; Poesen, J. A data-based landslide susceptibility map of Africa. *Earth-Sci. Rev.* **2018**, *185*, 102–121. [CrossRef]
43. Depicker, A.; Jacobs, L.; Delvaux, D.; Havenith, H.-B.; Maki Mateso, J.-C.; Govers, G.; Dewitte, O. The added value of a regional landslide susceptibility assessment: The western branch of the East African Rift. *Geomorphology* **2020**, *353*, 106886. [CrossRef]
44. Abalos, B. Geologic Map of the Basque Cantabrian Basin and a new tectonic interpretation of the Basque Arc. *Int. J. Earth Sci.* **2016**, *105*, 2327–2354. [CrossRef]
45. Cox, R.; Zentner, D.B.; Kirchner, B.J.; Cook, M.S. Boulder ridges on the Aran Islands (Ireland): Recent movements caused by storm waves, not tsunamis. *J. Geol.* **2012**, *120*, 249–272. [CrossRef]
46. Hall, A.M.; Hansom, J.D.; Jarvis, J. Patterns and rates of erosion produced by high energy wave processes on hard rock headlands: The Grind of the Navir, Shetland, Scotland. *Mar. Geol.* **2008**, *248*, 28–46. [CrossRef]
47. Hall, A.M.; Hansom, J.D.; Williams, D.M.; Jarvis, J. Distribution, geomorphology and lithofacies of cliff-top storm deposits: Examples from the high-energy coasts of Scotland and Ireland. *Mar. Geol.* **2006**, *232*, 131–155. [CrossRef]
48. Hastewell, L.J.; Schaefer, M.; Bray, M.; Inkpen, R. Intertidal boulder transport: A proposed methodology adopting Radio Frequency Identification (RFID) technology to quantify storm induced boulder mobility. *Earth Surf. Process. Landf.* **2019**, *44*, 681–698. [CrossRef]
49. Herterich, J.G.; Cox, R.; Dias, F. How does wave impact generate large boulders? Modelling hydraulic fracture of cliffs and shore platforms. *Mar. Geol.* **2018**, *339*, 34–46. [CrossRef]
50. McKenna, J.; Jackson, D.W.T.; Cooper, J.A.G. In situ exhumation from bedrock of large rounded boulders at the Giant's Causeway, Northern Ireland: An alternative genesis for large shore boulders (mega-clasts). *Mar. Geol.* **2011**, *283*, 25–35. [CrossRef]
51. Sommerville, A.A.; Hansom, J.D.; Sanderson, D.C.W.; Housley, R.A. Optically stimulated luminescence dating of large storm events in Northern Scotland. *Quat. Sci. Rev.* **2003**, *22*, 1085–1092. [CrossRef]
52. Autret, R.; Dodet, G.; Suanez, S.; Roudaut, G.; Fichaut, B. Long–term variability of supratidal coastal boulder activation in Brittany (France). *Geomorphology* **2018**, *304*, 184–200. [CrossRef]
53. Fichaut, B.; Suanez, S. Quarrying, transport and deposition of cliff-top storm deposits during extreme events: Banneg Island, Brittany. *Mar. Geol.* **2011**, *283*, 36–55. [CrossRef]
54. Pierre, G. Processes and rate of retreat of the clay and sandstone sea cliffs of the northern Boulonnais (France). *Geomorphology* **2006**, *73*, 64–77. [CrossRef]
55. Regnauld, H.; Oszwald, J.; Planchon, O.; Pignatelli, C.; Piscitelli, A.; Mastronuzzi, G.; Audevard, A. Polygenetic (tsunami and storm) deposits? A case study from Ushant Island, western France. *Z. Geomorphol.* **2010**, *54*, 197–217. [CrossRef]

56. Suanez, S.; Fichaut, B.; Magne, R. Cliff-top storm deposits on Banneg Island, Brittany, France: Effects of giant waves in the Eastern Atlantic Ocean. *Sediment. Geol.* **2009**, *220*, 12–28. [CrossRef]

57. Costa, P.J.M.; Andrade, C.; Freitas, M.C.; Oliveira, M.A.; da Silva, C.M.; Omira, R.; Taborda, R.; Baptista, M.A.; Dawson, A.G. Boulder deposition during major tsunami events. *Earth Surf. Process. Landf.* **2011**, *36*, 2054–2068. [CrossRef]

58. Horacio, J.; Munoz-Narciso, E.; Trenhaile, A.S.; Perez-Alberti, A. Remote sensing monitoring of a coastal-valley earthflow in northwestern Galicia, Spain. *Catena* **2019**, *178*, 276–287. [CrossRef]

59. Kortekaas, S.; Dawson, A.G. Distinguishing tsunami and storm deposits: An example from Martinhal, SW Portugal. *Sediment. Geol.* **2007**, *200*, 208–221. [CrossRef]

60. Oliveira, M.A.; Andrade, C.; Freitas, M.C.; Costa, P.; Taborda, R.; Janardo, C.; Neves, R. Transport of large boulders quarried from shore platforms of the Portuguese west coast. *J. Coast. Res.* **2011**, *64*, 1871–1875.

61. Ruiz, F.; Rodriguez-Ramirez, A.; Caceres, L.M.; Vidal, J.R.; Carretero, M.I.; Abad, M.; Olias, M.; Pozo, M. Evidence of high-energy events in the geological record: Mid-Holocene evolution of the southwestern Donana National Park (SW Spain). *Palaeogeogr. Palaeoclimatol. Palaeoecol.* **2005**, *229*, 212–229. [CrossRef]

62. European Severe Weather Database. Available online: https://www.eswd.eu (accessed on 6 February 2020).

63. Aubourg, C.; Giordano, G.; Mattei, M.; Speranza, F. Magma flow in sub-aqueous rhyolitic dikes inferred from magnetic fabric analysis (Ponza Island, W. Italy). *Phys. Chem. Earth* **2002**, *27*, 1263–1272. [CrossRef]

64. Bellucci, F.; Grimaldi, M.; Lirer, L.; Rapolla, A. Structure and geological evolution of the island of Ponza, Italy: Inferences from geological and gravimetric data. *J. Volcanol. Geotherm. Res.* **1997**, *79*, 87–96. [CrossRef]

65. DeRita, D.; Giordano, G.; Cecili, A. A model of submarine rhyolite dome growth: Ponza Island (central Italy). *J. Volcanol. Geotherm. Res.* **2001**, *107*, 221–239. [CrossRef]

66. Carobene, L.; Cevasco, A.; Firpo, M. Aspects of the Quaternary evolution of the coast between Cogoleto and Varazze (western Liguria). *Alp. Mediterr. Quat.* **2010**, *23*, 163–180.

67. Pepe, F.; Corradino, M.; Parrino, N.; Besio, G.; Presti, V.L.; Renda, P.; Calcagnile, L.; Quarta, G.; Sulli, A.; Antonioli, F. Boulder coastal deposits at Favignana Island rocky coast (Sicily, Italy): Litho-structural and hydrodynamic control. *Geomorphology* **2018**, *303*, 191–209. [CrossRef]

68. Piscitelli, A.; Milella, M.; Hippolyte, J.-C.; Shah-Hosseini, M.; Morhange, C.; Mastronuzzi, G. Numerical approach to the study of coastal boulders: The case of Martigues, Marseille, France. *Quat. Int.* **2017**, *439*, 52–64. [CrossRef]

69. Scotney, P.M.; Roberts, S.; Herrington, R.J.; Boyce, A.J.; Burgess, R. The development of volcanic massive sulfide and barite-gold orebodies on Wetar Island, Indonesia. *Miner. Depos.* **2005**, *40*, 76–99. [CrossRef]

70. Boesl, F.; Engel, M.; Eco, R.C.; Galang, J.B.; Gonzalo, L.A.; Llanes, F.; Quix, E.; Bruckner, H. Digital mapping of coastal boulders—High-resolution data acquisition to infer past and recent transport dynamics. *Sedimentology* **2019**, in press. [CrossRef]

71. Etienne, S.; Buckley, M.; Paris, R.; Nandasena, A.K.; Clark, K.; Strotz, L.; Chague-Goff, C.; Goff, J.; Richmond, B. The use of boulders for characterising past tsunamis: Lessons from the 2004 Indian Ocean and 2009 South Pacific tsunamis. *Earth-Sci. Rev.* **2011**, *107*, 76–90. [CrossRef]

72. Feldens, P.; Schwarzer, K.; Sakuna, D.; Szczucinski, W.; Sompongchaiyakul, P. Sediment distribution on the inner continental shelf off Khao Lak (Thailand) after the 2004 Indian Ocean tsunami. *Earth Planets Space* **2012**, *64*, 875–887. [CrossRef]

73. Fujino, S.; Sieh, K.; Meltzner, A.J.; Yulianto, E.; Chiang, H.-W. Ambiguous correlation of precisely dated coral detritus with the tsunamis of 1861 and 1907 at Simeulue Island, Aceh Province, Indonesia. *Mar. Geol.* **2014**, *357*, 384–391. [CrossRef]

74. Goto, K.; Chavanich, S.A.; Imamura, F.; Kunthasap, P.; Matsui, T.; Minoura, K.; Sugawara, D.; Yanagisawa, H. Distribution, origin and transport process of boulders deposited by the 2004 Indian Ocean tsunami at Pakarang Cape, Thailand. *Sediment. Geol.* **2007**, *202*, 821–837. [CrossRef]

75. Goto, K.; Okada, K.; Imamura, F. Numerical analysis of boulder transport by the 2004 Indian Ocean tsunami at Pakarang Cape, Thailand. *Mar. Geol.* **2010**, *26*, 97–105. [CrossRef]

76. Haslett, S.K.; Wong, B.R. An evaluation of boulder deposits along a granite coast affected by the 2004 Indian Ocean tsunami using revised hydrodynamic equations: Batu Ferringhi, Penang, Malaysia. *J. Geol.* **2019**, *127*, 527–541. [CrossRef]

77. Kennedy, A.B.; Mori, N.; Yasuda, T.; Shimozono, T.; Tomiczek, T.; Donahue, A.; Shimura, T.; Imai, Y. Extreme block and boulder transport along a cliffed coastline (Calicoan Island, Philippines) during Super Typhoon Haiyan. *Mar. Geol.* **2017**, *383*, 65–77. [CrossRef]

78. Kennedy, A.B.; Mori, N.; Zhang, Y.; Yasuda, T.; Chen, S.-E.; Tajima, Y.; Pecor, W.; Toride, K. Observations and Modeling of Coastal Boulder Transport and Loading during Super Typhoon Haiyan. *Coast. Eng. J.* **2016**, *58*, 1640004. [CrossRef]

79. Lau, A.Y.A.; Terry, J.P.; Switzer, A.D.; Pile, J. Advantages of beachrock slabs for interpreting high-energy wave transport: Evidence from Ludao island in south-eastern Taiwan. *Geomorphology* **2015**, *228*, 263–274. [CrossRef]

80. May, S.M.; Engel, M.; Brill, D.; Cuadra, C.; Lagmay, A.M.F.; Santiago, J.; Suarez, J.K.; Reyes, M.; Bruckner, H. Block and boulder transport in Eastern Samar (Philippines) during Supertyphoon Haiyan. *Earth Surf. Dyn.* **2015**, *3*, 543–558. [CrossRef]

81. Paris, R.; Fournier, J.; Poizot, E.; Etienne, S.; Morin, J.; Lavigne, F.; Wassmer, P. Boulder and fine sediment transport and deposition by the 2004 tsunami in Lhok Nga (western Banda Aceh, Sumatra, Indonesia): A coupled offshore-onshore model. *Mar. Geol.* **2010**, *268*, 43–54. [CrossRef]

82. Paris, R.; Wassmer, P.; Sartohadi, J.; Lavigne, F.; Barthomeuf, B.; Desgages, E.; Grancher, D.; Baumert, P.; Vautier, F.; Brunstein, D.; et al. Tsunamis as geomorphic crises: Lessons from the December 26, 2004 tsunami in Lhok Nga, West Banda Aceh (Sumatra, Indonesia). *Geomorphology* **2009**, *104*, 59–72. [CrossRef]

83. Scheucher, L.E.A.; Vortisch, W. Sedimentological and geomorphological effects of the Sumatra-Andaman Tsunami in the area of Khao Lak, southern Thailand. *Environ. Earth Sci.* **2011**, *63*, 785–796. [CrossRef]

84. Soria, J.L.A.; Switzer, A.D.; Pilarczyk, J.E.; Tang, H.; Weiss, R.; Siringan, F.; Manglicmot, M.; Gallentes, A.; Lau, A.Y.A.; Cheong, A.Y.L.; et al. Surf beat-induced overwash during Typhoon Haiyan deposited two distinct sediment assemblages on the carbonate coast of Hernani, Samar, central Philippines. *Mar. Geol.* **2018**, *396*, 215–230. [CrossRef]

85. Szczucinski, W. The post-depositional changes of the onshore 2004 tsunami deposits on the Andaman Sea coast of Thailand. *Nat. Hazards* **2012**, *60*, 115–133. [CrossRef]

86. Terry, J.P.; Dunne, K.; Jankaew, K. Prehistorical frequency of high-energy marine inundation events driven by typhoons in the Bay of Bangkok (Thailand), interpreted from coastal carbonate boulders. *Earth Surf. Process. Landf.* **2016**, *41*, 553–562. [CrossRef]

87. Terry, J.P.; Goff, J. Strongly aligned coastal boulders on Ko Larn island (Thailand): A proxy for past typhoon-driven high-energy wave events in the Bay of Bangkok. *Geogr. Res.* **2019**, *57*, 344–358. [CrossRef]

88. Terry, J.P.; Goff, J.; Jankaew, K. Major typhoon phases in the upper Gulf of Thailand over the last 1.5 millennia, determined from coastal deposits on rock islands. *Quat. Int.* **2018**, *487*, 87–98. [CrossRef]

89. Terry, J.P.; Jankaew, K.; Dunne, K. Coastal vulnerability to typhoon inundation in the Bay of Bangkok, Thailand? Evidence from carbonate boulder deposits on Ko Larn island. *Estuar. Coast. Shelf Sci.* **2015**, *165*, 261–269. [CrossRef]

90. Terry, J.P.; Oliver, G.J.H.; Friess, D.A. Ancient high-energy storm boulder deposits on Ko Samui, Thailand, and their significance for identifying coastal hazard risk. *Palaeogeogr. Palaeoclimatol. Palaeoecol.* **2016**, *454*, 282–293. [CrossRef]

91. Xu, X.; Gao, S.; Zhou, L.; Yang, B. Sedimentary records of extreme wave events on the northeastern Hainan Island coast, southern China. *Haiyang Xuebao* **2019**, *41*, 48–63.

92. Yang, W.; Sun, L.; Yang, Z.; Gao, S.; Gao, Y.; Shao, D.; Mei, Y.; Zang, J.; Wang, Y.; Xie, Z. Nan'ao, an archaeological site of Song dynasty destroyed by tsunami. *Chin. Sci. Bull.* **2019**, *64*, 107–120. [CrossRef]

93. Yawsangratt, S.; Szczucinski, W.; Chaimanee, N.; Jagodzinski, R.; Lorenc, S.; Chatprasert, S.; Saisuttichai, D.; Tepsuwan, T. Depositional effects of 2004 tsunami and hypothetical paleotsunami near Thap Lamu Navy Base in Phang Nga Province, Thailand. *Pol. J. Environ. Stud.* **2009**, *18*, 17–23.

94. Zhou, L.; Gao, S.; Jia, J.; Zhang, Y.; Yang, Y.; Mao, L.; Fang, X.; Shulmeister, J. Extracting historic cyclone data from coastal dune deposits in eastern Hainan Island, China. *Sediment. Geol.* **2019**, *392*, 105524. [CrossRef]

95. Gusiakov, V.K.; Dunbar, P.K.; Arcos, N. Twenty-Five Years (1992–2016) of Global Tsunamis: Statistical and Analytical Overview. *Pure Appl. Geophys.* **2019**, *176*, 2795–2807. [CrossRef]

96. Løvholt, F.; Glimsdal, S.; Harbitz, C.B.; Zamora, N.; Nadim, F.; Peduzzi, P.; Dao, H.; Smebye, H. Tsunami hazard and exposure on the global scale. *Earth-Sci. Rev.* **2012**, *110*, 58–73. [CrossRef]

97.  Kang, N.-Y.; Kim, D.; Elsner, J.B. The contribution of super typhoons to tropical cyclone activity in response to ENSO. *Sci. Rep.* **2019**, *9*, 5046. [CrossRef]
98.  Barat, F.; Mercier de Lépinay, B.; Sosson, M.; Müller, C.; Baumgartner, P.O.; Baumgartner-Mora, C. Transition from the Farallon Plate subduction to the collision between South and Central America: Geological evolution of the Panama Isthmus. *Tectonophysics* **2014**, *622*, 145–167. [CrossRef]
99.  Bahlburg, H.; Spiske, M. Sedimentology of tsunami inflow and backflow deposits: Key differences revealed in a modern example. *Sedimentology* **2012**, *59*, 1063–1086. [CrossRef]
100. Bahlburg, H.; Nentwig, V.; Kreutzer, M. The September 16, 2015 Illapel tsunami, Chile—Sedimentology of tsunami deposits at the beaches of La Serena and Coquimbo. *Mar. Geol.* **2018**, *396*, 43–53. [CrossRef]
101. Evers, C.; Doering, A. Lifestyle Sports in East Asia. *J. Sport Soc. Issues* **2019**, *43*, 343–352. [CrossRef]
102. Magiera, A.; Roczniok, R. The climbing preferences of advanced rock climbers. *Hum. Mov.* **2013**, *14*, 254–264. [CrossRef]
103. Schwartz, F.; Taff, B.D.; Lawhon, B.; Pettebone, D.; Esser, S.; D'Antonio, A. Leave No Trace bouldering ethics: Transitioning from the gym to the crag. *J. Outdoor Recreat. Tour.* **2019**, *25*, 16–23. [CrossRef]
104. Tessler, M.; Clark, T.A. The impact of bouldering on rock-associated vegetation. *Biol. Conserv.* **2016** *204*, 426–433. [CrossRef]
105. Van der Merwe, J.H.; Joubert, U. Managing environmental impact of bouldering as a niche outdoor-climbing activity. *S. Afr. J. Res. Sport, Phys. Educ. Recreat.* **2014**, *36*, 229–251.
106. 27 Crags. Available online: https://27crags.com/ (accessed on 6 February 2020).
107. Climb Europe. Available online: https://www.climb-europe.com/EuropeanRockClimbingAreas.html (accessed on 6 February 2020).
108. Climbing Away. Available online: https://climbingaway.fr/en/rock-climbing-areas/world-map-of-rock-climbing-areas (accessed on 6 February 2020).
109. Liu, Z.Y.-C.; Harris, R.A. Discovery of possible mega-thrust earthquake along the Seram Trough from records of 1629 tsunami in eastern Indonesian region. *Nat. Hazards* **2014**, *72*, 1311–1328. [CrossRef]
110. Tinti, S.; Maramai, A.; Graziani, L. A new version of the European tsunami catalogue: Updating and revision. *Nat. Hazards Earth Syst. Sci.* **2001**, *1*, 255–262. [CrossRef]
111. Khanal, N.; Uddin, K.; Matin, M.A.; Tenneson, K. Automatic detection of spatiotemporal urban expansion patterns by fusing OSM and Landsat data in Kathmandu. *Remote Sens.* **2019**, *11*, 2296. [CrossRef]
112. Liss, B.; Howland, M.D.; Levy, T.E. Testing Google Earth Engine for the automatic identification and vectorization of archaeological features: A case study from Faynan, Jordan. *J. Archaeol. Sci. Rep.* **2017**, *15*, 299–304. [CrossRef]
113. Luijendijk, A.; Hagenaars, G.; Ranasinghe, R.; Baart, F.; Donchyts, G.; Aarninkhof, S. The State of the World's Beaches. *Sci. Rep.* **2018**, *8*, 6641. [CrossRef]
114. Vos, K.; Splinter, K.D.; Harley, M.D.; Simmons, J.A.; Turner, I.L. CoastSat: A Google Earth Engine-enabled Python toolkit to extract shorelines from publicly available satellite imagery. *Environ. Model. Softw.* **2019**, *122*, 104528. [CrossRef]
115. Moreno Chávez, G.; Sarocchi, D.; Arce Santana, E.; Borselli, L.; Rodríguez-Sedano, L.A. Using Kinect to analyze pebble to block-sized clasts in sedimentology. *Comput. Geosci.* **2014**, *72*, 18–32. [CrossRef]
116. Říha, K.; Krupka, A.; Costa, P.J.M. Image analysis applied to quartz grain microtextural provenance studies. *Comput. Geosci.* **2019**, *125*, 98–108. [CrossRef]
117. Tunwal, M.; Mulchrone, K.F.; Meere, P.A. Image based Particle Shape Analysis Toolbox (IPSAT). *Comput. Geosci.* **2020**, *135*, 104391. [CrossRef]

MDPI

St. Alban-Anlage 66

4052 Basel

Switzerland

Tel. +41 61 683 77 34

Fax +41 61 302 89 18

www.mdpi.com

*Journal of Marine Science and Engineering* Editorial Office

E-mail: jmse@mdpi.com

www.mdpi.com/journal/jmse